...What Helicopter People Do For People

LIFT

IS WHERE YOU FIND IT

Joe Stein

The Zig Zag Papers

Box 247 Zig Zag, Oregon 97049

Photo Credits

Page 3,4 National Air Museum; 16, 17 U.S. Coast Guard; 25 U. S. Park Police; 29 Steve Frederick; 31 Univ. Miss. Medical Center; 32-upper Emanuel Hosp.; 32-lower Washington Hosp. Center; 47 Joe Melena; 59-upper Evergreen Helicopters; 59-lower Carrol Voss, Ag Rotors; 67 Ted Veal, Columbia Helicopters; 71-upper Houston Police Dept.; 71-lower Maryland State Police; 86 Bell Helicopter Textron; 91 British Air Helicopters; 101, 112 Ralph Perry; 114, 123 Ted Veal, Columbia; 119 Aerospatiale Helicopter Corp.; 127-upper Bonneville Power Admin.; 127-lower Ted Veal, Columbia; 147 Alfred A. Monner; 148 MBB Helicopter Corp.; 149, 158 Sikorsky Aircraft; 164, 165-upper U. S. Army; 165-lower Aerospatiale; 170, 177 Sikorsky Aircraft; 180 Bell Helicopter; 191-upper Sikorsky Aircraft; 191-lower Bell Helicopter; 193 Robinson Helicopter; 206 Ted Veal; 207 London Metro Police.

Catalog data — 1.aerial cranes, 2.aeronautics, history 3.agriculture, 4.air transport, 5.aviation, military, 6.aviation, sport, 7.electric power lines, 8.electronic news reporting, 9.firefighting, 10.forestry, 11.helicopters, 12.law enforcement, 13.medicine, emergency, 14.oil industry, 15.pest control, 16.physics of flight, 17.Presidents.

Edited by *Pam Horan and Merle Pugh*
Front Cover Painting by *Nancy Stein*
Back Cover Color Photo by *Ralph Perry*
Drawings by *Dean McMullen*
Calligraphy by *Carol Campbell*
Book Design by *Joanne Stevens Sullivan*

First Edition
Revised, Second Printing May 1986
Printed by Catalogs Unlimited
United States of America

Library of Congress Catalogue Number 85-90111

ISBN 0-9614498-3-7

DEDICATION

To the pilots and the mechanics, to Sam and Phil Jackson: time has affirmed your faith.

ACKNOWLEDGEMENTS

As nothing lifts like a helicopter, it is so with helicopter people. They **lift.** Their spirit contributed to every page of this book. I owe special thanks to Speed Williams, Fred Snell, Ralph Perry, Bill Snyder, Carrol Voss, John Ward, Alfred Gessow, and Jock Cameron for ready and constant encouragement.

Real Lift, too, came from Al Arnason, Bill Caraway, Butch Cronin, Brad Dunbar, Vic Hefferin, Terry Jagerson, Frank Jensen, Richard M. Long, Ted Millar, Bob Morrison, Ed Mosey, Chuck Nole, Allen Price, Ed Pope, Marty Reisch, Myles Ruggenberg, Del Smith, Bill Thomas, Ted Veal, and at alphabet's end, John Zugschwert and John Zuk.

Contributing directly were my daughter, Nancy — front cover water color; Ralph Perry — back cover photo of Coldwater Ridge, Washington; my daughter, Carol Campbell — calligraphy; Dean McMullen — logo and drawings; Pam Horan and Merle Pugh — editing and editorial advice; Joanne Stevens Sullivan — design; Dean and Marlys Burn, and Dolores Green — typesetting, pasteup and printing; my wife, Jean — wise and generous counsel, and for inspiring our artist daughters.

Contents

Chapter

Foreword

Joe Stein has given us a different helicopter book. There are excellent books about pioneers such as Igor Sikorsky and Lawrence Bell. There are many books about the early development of the helicopter. Some recent books capture the feel of helicopters in combat. But I know of no book that tells so well the story of the helicopter performing many tasks successfully throughout the world today.

Some helicopter tasks, such as its role in rescues at sea and on land are well-known. But many are unknown, and they even surprise helicopter pilots.

Stein tells us about this multitude of uses, and he does so through the first-hand accounts of the men and women who were involved. Through their stories we feel and understand the problems and difficulties which have been overcome in the helicopter, and we understand those that persist, a few of which may well prove to be insurmountable. Indeed, they have long defied solution.

This book helps us understand and appreciate the qualities of the "new pioneers" of our own time — their imagination, persistence, and an unwavering belief in their own ability and the capabilities of a machine that was the butt of jokes throughout a large part of the aviation fraternity not many years ago. Between the lines we see that the helicopter, ancient in concept and old in action after forty years of general use, is as new and exciting as tomorrow.

Whether you read this book for knowledge or pleasure, it will give you both. On a personal note, I have cherished the friendship of Joe Stein for more than thirty-five years — a fellow helicopter pilot in the early days, and that alone distinguishes him from the run of aviators. He is one of those who, through World War II and after, flew professionally because of deep conviction and enthusiasm.

My friend Joe Stein has done well as a writer of aviation and science, and on the broad subject of humanity. His love of people, of copters and of flight is apparent throughout this book.

M. B. "Speed" Williams
Captain, USCG, retired
Potomac, Maryland

Introduction: Free Wings

You can go anywhere in a helicopter. It flies straight, direct from point to point, not just to the airport outside of town. It has that unique go-anywhere character because rotor-blade wings turn free in the air without otherwise moving. They create lift, and lift gives travel a vertical dimension.

But multitudes don't go anywhere by helicopter, and there's our story. High costs and lack of convenient access curb this versatile vehicle. The skeptics decry its low speed, its evident hazards and noise as well.

It is merely the plaything of the jet set? A contraption? Why so expensive? How real the defects, the dangers of the funny gadget we see on TV? The fixed wing airplane faced skepticism too, not long ago. Depending on your perspective, it was costly or impractical or slow, compared to what?

Airplanes can glide and soar like the birds, but the likeness ends there. Birds need no wheels nor floats nor skis to leave the Earth and return. For the fixed wing we have built miles and miles of runways and choked terminals too — so we must travel to and from large airports to use its obvious benefits.

Since rigid wings cannot fly without moving forward, they must land where the airport is. With free-turning wings, copters can fly slow or stop, high or low; they demand no special facilities — and so, more like the birds and bees with that vertical dimension. They are unique among vehicles with *lift anywhere.*

Our ancestors saw birds fly everywhere at will. What could be more free? For centuries they tried but found no way to emulate the birds' freedom of motion. Such was their dream. Speed came later as a goal in the evolution of flight.

When at last the fixed wing airplane flew, it went on to speeds unimagined, and began shrinking the Earth. Now, as aviation's first century nears its close, the airplane dominates the sky in warfare and long-distance travel. Still tied to airports, however, it cannot fully answer The Dream.

Early inventors tried free rotating wings, but found fixed wings easier to understand and simpler to engineer. No wonder they made better progress with the airplane. In three decades the complex helicopter evolved by slow degrees, from possible to primitive to probable to promising to practical. Today it still lags the fixed wing by thirty years.

Slow, yes, but a sure evolution. Free-rotating wings have no equal in rescue or patrol or the unique "sling load." As it alone lifts straight up, the helicopter does certain tasks better than any other vehicle. It works where people live and do business; it can be the best way to travel distances up to 600 miles. Nothing *lifts* like a helicopter.

We know much about helicopter failures. Bold headlines told of tragic losses in Viet Nam: disaster in attempted release of American hostages in Iran; crashes upon building roofs in large cities. The same history recounts prodigious rescue on volcanos; in hotel and casino fires; airliner crash in the heart of Washington; sinking ships, storms, and floods. And, quiet but sure for almost thirty years, Presidents have begun local travels and distant ones, from the White House lawn.

When critics find this colorful record "mixed," they ignore the carnage of highways, railways, airways, the constant risks of travel — even while traffic congestion grows everywhere, along with inherent dangers. The helicopter has the unique safe landing feature of autorotation[1].

It is an old story in human history; it was so with sailing ships, with canal boats, trains, and autos. And when airlines were few and crashes more novel, few passengers dared the small planes. They paid fares three, four, ten times as high as by train or bus, and everybody *knew* that flying was dangerous. As airplanes grew in size, went faster, had more seats and comforts *at lower fares,* people found them safer, somehow.

High costs squelch a wider popularity for helicopters. With only limited production, they are virtually hand-crafted, and so, priced high. In 1984 the smallest available listed at $82,000. Still the Dream persists, to fly like birds, with free vertical lift.

In the four decades of progress with helicopters, human ingenuity has reduced the cost while improving dependability and performance. Today some 10,000 civilian helicopters fly American skies; they total over 16,000 around the world. To be sure, they serve humanity out of proportion to small numbers.

Thousands more serve the fighting forces. The first crude designs went direct to the crushing demands of warfare. The copter has superseded the military seaplane; it continues finding new utility with Navies and Marines. And soldiers insist it is a deadly killer of tanks.

Helicopters and fixed wings compete only in part. Both

[1] See appendix.

carry large loads and go high, both fly by the laws of physics. But each has its own place, its mode. The airplane found its role in supreme speed. With vertical flight, the helicopter fills a palpable void between the lofty airway and the ground-bound vehicle. To mix an old metaphor, it beats a shorter path to your door.

Like other machines it exists for human benefit. So who is served? Its beneficiaries are people such as depicted in this book, in stories of what the copter means to a world of pressing needs.

Those who dreamed The Dream risked life and limb and fortune, and frequently lost for all their pains. A few succeeded by persistence, with whatever they had. A world bedazzled with fighter jockeys and supersonic flight gave but sparse interest, financial aid, and support.

Still, the helicopter drew its own loyal adherents. In the words of Alfred Gessow, University of Maryland professor, rotor wing scientist and author of books and papers, "People working in this field have an emotional attachment you never find with fixed wings. It is far more to them than a business or an engineering job."

The helicopter is a metaphor for the human dream of freedom.

Among The Dreamers were the uneducated and those too unwise to heed the venerable "truths" of the past. That was fortunate, for the so-called truths often serve to deny human progress. The fixed wing zealots who dote on speed are no wiser than the rotor-wing promoters who would ignore whatever impedes helicopter progress.

Of those troubles, two resist solution, high price and convenient access. It once was so with the airplane, whose relative costs have come down over the decades; despite the burden of expense, we paid to develop a truly immense complex of facilities — airports, terminals, airways, and ground fixtures for fixed wings. Now fixed wing evolution has reached a plateau. What can further development gain for it in speed or power or efficiency? Will we build what we need for helicopters?

Although modest in demand, heliports meet stiff opposition in urban places. The copter poses a question, a threat, and a promise for its future. Its long past record leads to the question of cost and efficiency — to the threat that its difficult problems may not yield to solution — yet it has that promise of realizing the potential utility of free wings.

That history, four decades of slow but real success, attests that the best may take a little longer.

A Fable

When Helo was young, Father Igor warned, "You must work. You must surpass Sky King who dominates this world." And Helo toiled with exceeding zeal.

Helo beheld Sky King who soared over the Very Top. He esteemed his large and lordly wings, his Speed-That-Transcends-All.

But verily, burdened with rigid wings Sky King flew not unto Perfection. Ungainly at the ground was he, without hover, who indeed must strain at Gravity. To aviate he must run Far on Solid Earth, consort with jet streams on high, and must not but alight on Broad Pastures. To dale, islet, crag; to dock, roof, street, and parking lot, to Myriad Places In Between, he ventured not. Forsooth, he bore the burden of Speed-That-Transcends-All.

In wisdom Helo made slow progress, yet he braved Gravity in all climes. Sore beset with White Knuckles and vibration, he did grow intimate with rain and snow, ice and fog — indeed with wires too— and trees, rocks, grass, and waters, in Nature's bounty, Down Here among living things. Where only God can make a Tree.

It came to pass, Helo won the hearts of Pharmers and Persons in Peril, and doubting hosts. He carried parcels, pots, poets, petroleum parts, politicos, popsicles, popes, pythons, pandas, pretzels, pets, policemen, paleontologists, photogs, pharmacists, Phillippinos, and Philistines, PM's POW's, PFC's and prefects; postal pouches, purines, Presidents, power poles, patients, potentates, and ponderosa pines. Also Chancellors, Chairmen of the Board, and beadles betimes.

To land's end and beyond he served their aspirations, in profuse tasks and places unavailing to Sky King. The multitudes besought our Helo in war and peace and Otherwise Elsewhere. Igor did rejoice to see his diffusing fame, and Gravity did submit to his swirling wings.

For it is written: *A straight flight is the shortest distance unto the human heart.*

1.

The President's Chariot

President Eisenhower's helicopter was a good machine. I think it was made by Sikorsky.

- Nikita Kruschchev

For almost three decades American Presidents have been riding helicopters in comfort and style. Seven Presidents in a row have had door-to-door service and privacy. Seven Presidents since 1957 have traveled safe in the air and secure from assassins.

All through history Presidents have traveled in the mode of their day, from the horse-drawn carriage to the long black limousine. In Niagara Falls, New York they introduced President Harry Truman to the world's first civil helicopter, but he did not ride one until he retired. The Secret Service would not accept the helicopter; the guards had worries enough with Truman's frequent walks in the streets of Washington.

After Truman, walking was out. Jimmy Carter took a memorable hike, from his 1977 inauguration at the Capitol to his White House home. That 2-mile jaunt on Pennsylvania Avenue gave his guards the horrors. Thereafter they sequestered him in cars and copters.

Although secrecy shrouds the chief executive's travel habits, everybody knows he rides in bullet-proof cars with bullet-proof glass, shielded by bullet-proofed guards. It's the same for his helicopters, all armor plated and safetied to the limit of technology.

Cargo jets carry both copters and cars to distant places, for use on local junkets. Helicopters can, if necessary, fly the President 500 miles without refueling.

Unsurprisingly, each President has made the most of this rapid air transit in his own way. But in surprising fact, the presidency has kept the helicopter its virtual exclusive in the National Capital.

Since aviation's earliest days, laws have prohibited flying over the city — a Prohibited Air Space covering the White House, the Capitol, and everything between. That zone remains sacrosanct for copters of the President and a few others in the federal institution. Does it smack of the "imperial presidency" as critics say? Or does security justify for this popular target of assassins?

The helicopter arrived late, July 12, 1957, but promptly became a fixture, alongside the black cars and airplanes of the White House. In his Army career, President Dwight Eisenhower knew airplanes and a little of primitive whirly wings. Helicopters already had fifteen years of use, in warfare at first, then in civil hands, again in the Korean war. But like the airplane of three decades before, it had but little popularity at that time.

First Presidential Copter

The first presidential chariot, a Bell 47J Ranger (officially two USAF H13J's), had a bubble windshield and four seats, for one pilot up front, and three aft. A four-cylinder Lycoming engine of 260 horsepower turned the two-blade rotor. From tip to tail, it measured 32 ft. 4 in. and weighed 2,850 pounds at full load. It cruised at a rollicking 87 miles an hour — up and over all obstacles.

Fitting for Our Man, they painted the 47J red, white, and blue. Sure not plushy, maybe primitive, but as helicopters go it was small, and a real novelty in that day. Air Force pilots flew the President, who right away saw he could zip 66 air miles in 35 minutes to his farm at Gettysburg, Pennsylvania. Rather nicer than 90 minutes or more in the black auto on the state roads.

Bell Helicopter Company pilots Joe Mashman, Joe Dunne, and Harry Mitchell took a part in this aerial initiation. Mitchell and Dunne were Army veterans, and Mashman, a test pilot, helped to develop the 47J, at Fort Worth, Texas.

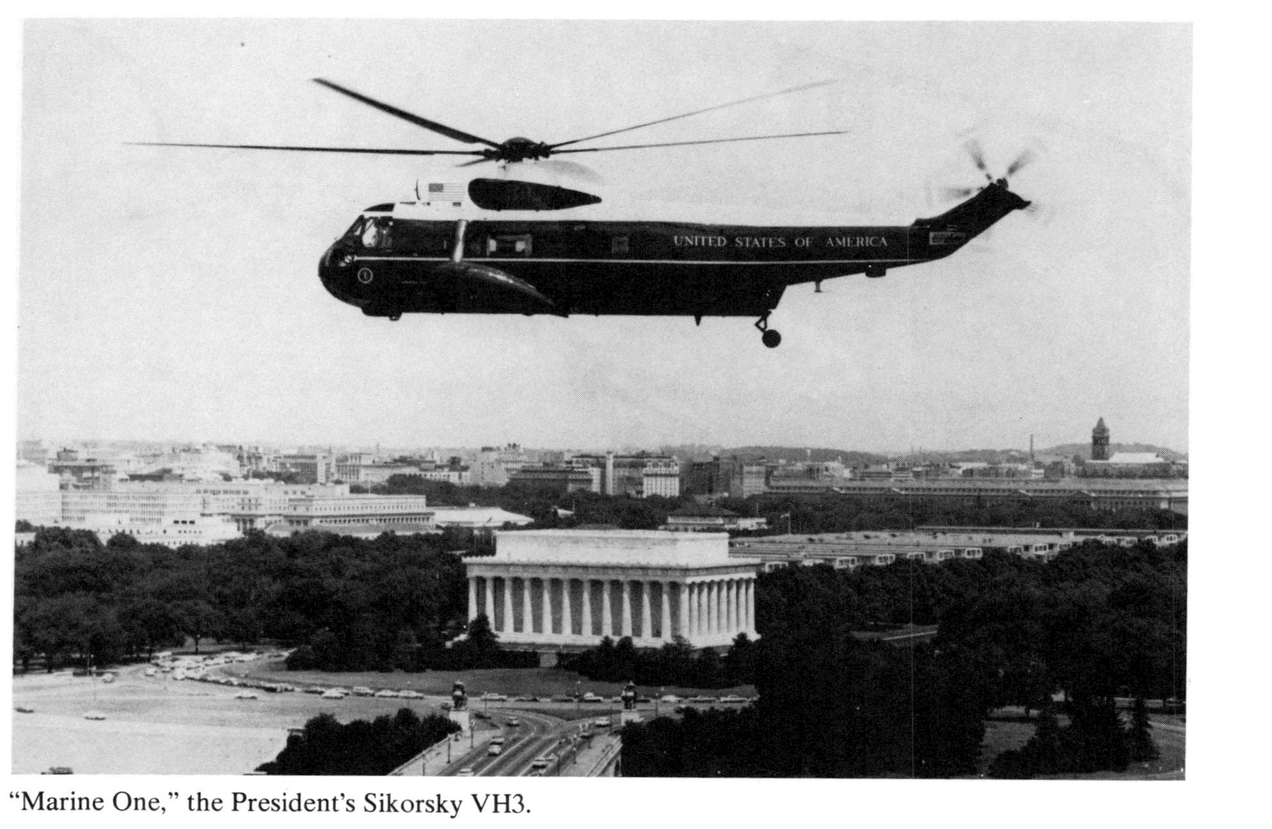

"Marine One," the President's Sikorsky VH3.

Bell 47J, first Presidential helicopter, 1957.

As Mitchell recounted it, "all of a sudden the boss hustled Dunne and me to Washington to look over the White House yard. He said, 'See if it's workable to land and take off there.' We went, quiet and fast. The President's son, John, furnished escorts and the run of that big yard. We made lots of notes — paced off the clear part, noted the trees, slope of the ground, the probable wind currents, the obstacles, the whole business — a piece of cake, as we saw right away.

"Of course, you could only approach from the south. Trees and buildings stand to east and west, and the house blocks the north. We asked ourselves, was it feasible to fly in and out the same way? You'd always fly down-wind, in or out, but that didn't matter. At worst you might have to make tight turns in there.

"For a helicopter, not a problem. Plenty of room to land and take off. And you know, for all the years since, the big ones have been doing just that, even the tight turns."

Mitchell and Dunne had no idea they were making history, just doing their job. "Who couldn't land a helicopter on that broad lawn? D. C. is full of wide, grassy places. Better than any other city I know. Why don't they use helicopters there?

"We were more impressed with the Washington Monument. It stands due south, in the approach path. We had no way to fly around the White House, so we could get a good overhead view from the Monument — 550 feet up and three blocks away. We walked up — 717 steps to the top of the Monument, isn't it? We remember that hike most of all. We were young, almost thirty years ago.

"Never went inside the White House; John Eisenhower didn't offer and neither of us thought of it. We went back to the plant and reported to Mashman. He flew it later — but I don't recall when."

Mashman does not remember the date either. It came just after Bell engineered the J version of the Model 47. He laughed over that start. "Management alloted $25,000 to make this four-seat model out of a three-seat 47G. Today that would cost hundreds of thousands. We cut off the bubble front and installed the larger one, put in four seats, and changed the controls. That was our prototype. It worked fine, and production went from there."

Mashman, Bell's dapper test pilot was well into his life-long career as diplomat-salesman. He and an engineer, Joe

Beebe, went on tour with the 47J. They showed it at a number of military bases, and spent three months in Central and South America.

"Early in January 1957 our Navy bought 47J's. (That order paid off development costs.) Then Eisenhower's Naval aide requested a demonstration. But first we had to get it past John Foster Dulles, the secretary of state. We flew it at his private island in Ontario.

"I was asked to land at the White House. Eisenhower came out to inspect it. He shook hands, sat in it, and asked questions. Later they ordered two as Air Force H13J's. Of course, they weren't content to let us make the finest we could for our President. Secret Service agents spent weeks at our plant. They passed on every screw and cotter key that went into it. They got two of the best helicopters Bell ever made."

The Air Force installed special equipment, and added more later. The first two helicopters cost the government $201,000. From the start, these copters flew always in pairs, carrying the Presidential party, guards, baggage, and paraphernalia.

Bigger Machine Needed

The 47J held favor only a short time, as the President found it too small and too slow. After two months it was superseded by the twelve-seat Sikorsky S58 (Army H34). The 56-ft, 8-in. S58 weighed 7,750 pounds, with a main rotor of 56-ft. diameter, cruising speed of 100 MPH, and a gasoline engine of 1,525 HP. The roomier S58 carried two pilots and an orderly who handed out pillows and cotton pads to dull some of the rotor racket. Presidents have flown Sikorsky's ever since.

Early in 1958, Marine Corps and Army pilots, who had more experience with rotorcraft, took over from the USAF. They alternated in teams, from week to week. In 1976, in Gerald Ford's term, the Marines of Quantico, Virginia, got the assignment fulltime, and the radio call sign became "Marine One."

Eisenhower took copters along on trips to Spain, Afghanistan, and Brazil. In 1959, with Premier Nikita Kruschchev, USSR, and several others, he toured Washington and flew to Camp David in Maryland. That junket impressed the Russian. Kruschchev wrote in his memoirs, "Eisenhower asked me if

I would mind flying to Camp David by helicopter. . . . The President's helicopter was a good machine. I think it was made by Sikorsky."

In President Kennedy's term the White House changed to the VH3, an extensively modified HH3 (civil S61). It has two engines, the finest of airliner accommodations, speed of 130 MPH, and low noise level. It seats as many as twenty passengers (on bench seats), or as few as ten with the President on board. It has a bar, a small galley, and a bathroom, but they get little use on its short hops. One or two enlisted Marines in dress uniform serve as steward-orderlies.

Variations of the VH3 have been in constant use since Kennedy's day. Powered by gas turbine (jet) engines of 1,500 HP, the VH3 has a 62-ft rotor of five blades and 10-ft. tail rotor; it measures 55 ft. 3 in. long and weighs 20,500 pounds fully loaded. Special metal and plastic materials protect occupants from hostile fire and malfunctioning machinery.

The close-mouthed White House guards such figures but informed sources say the VH3 costs "over $6 million new, and more than $3,000 an hour to operate." As a perquisite for the commander-in-chief, the money comes out of the Defense Department budget.

The latest model has dropdown step-doors at front and rear of the cabin. "About the only thing wrong with that," said a former pilot, "the doorway is low, under 6 feet. Kennedy, Johnson, Ford, and Reagan, the tall ones had to duck their heads as they entered."

At its Quantico, Virginia base, 40 miles south of D. C., Marine Corps Squadron HMX1 keeps a fleet — five different types of copters — for the President's needs. The squadron also does helicopter flight testing. A detachment of copters and crews stays at Anacostia, the former Naval Air Station located 3 miles from the White House.

The squadron maintains the various helicopter types with special features and fixtures (many of them secret) for the President, and spares as well. Crew members must live within 10 miles of Quantico — ready around the clock to be airborne within 20 minutes. The liaison office, usually the President's Marine aide, sends the orders. At the White House three concrete slabs serve as landing pads, a few steps from the back door.

"To us," a former Marine said, "it was an honor to serve, but more so, a heavy responsibility — let nothing ever happen to The Man. The idea is *security*. That helicopter flies there to get him out of harm's way no matter what!"

Pilots say President Kennedy, like the others, made full use of his copters. He played it loose, took his ease while riding, and could be careless about seat belts. "So we could order him around?" He always gave crews a thank you and a handshake, yet he had the air as well as the authority of commander-in-chief.

His wife ". . . didn't like Marines. We never knew why. She would lord it over us, but JFK took no notice. She'd carry sixty dresses for a trip. One time we took a load of hay to Camp David for her ponies."

"Both of them," he added, "let their kids play freely. We didn't mind. At times we hauled Marines to Camp David so the Kennedy's could play touch football. Didn't mind that either. They used the helos for anything they wanted. But that President – most Presidents made us feel good."

HMX crews undergo the most rigorous security clearances, normally taking eighteen months to complete. About fifty pilots are on staff; they are selected for skill and safety record in flying. They take extra training at the Navy's Pensacola, Florida, flight school. They are reticent about the experience; few will talk even off the record. All are men; so far the Marine Corps has no women pilots.

No Peer For Safety

On one point everybody agrees. Both for safety of flight and for security from attack, the helicopters have no peer. In both rotary and fixed wings, no President has had a bad moment, to the credit of the flying services. Every modern device and navigation aid is applied, and all maintenance – by both Marine and civil experts from Sikorsky – is done on the most conservative basis.

For example, a former pilot said, "they inspect rotor blades at 400 hours of service, and replace them at 1,200 hours. On the HH3 (basically the same as the VH3), we run blades 3,000 hours. They do everything that way for the President."

Presidents Johnson and Nixon made the most – and most lavish – use of helicopters, although Johnson posed as the frugal one. Former crew men said Mr. Johnson's wife and daughters were often boarded at the Pentagon or Anacostia, out of the White House limelight. Johnson called out more night flights than any other President — as late as 4:30 a.m.

Lyndon Johnson won the dislike of some pilots. "He was quick to give us hell," said one. "He'd storm up front yelling, 'What the fuck are you bastards up to now' when he was upset, and we didn't know what was wrong. Sure had a foul mouth, for a President."

Late in 1968, at the height of anti-Viet Nam war activity, Johnson summarily fired a pilot, Major Don Foss. It happened during a demonstration against the war. He announced he would ignore the opposition and would go to Camp David. Then, in secret, he stayed at the White House.

Foss, flying General William Westmoreland, the Viet Nam commander, had to cruise low over the Potomac River to stay clear of fixed wing traffic from Washington National Airport. That took the white-topped helicopter over the parade route. The demonstrators recognized it and assumed Johnson was spying on them, while he pretended to be out of town. They heaped scorn on him.

In a fury the President took swift action against the hapless pilot, and ordered Foss out of the squadron. But the Marine Corps gave him a good assignment when Johnson went out of office a few months later.

"Too many helicopters" always angered Johnson. One former pilot said he "ordered us not to bring up more than two choppers when he traveled. Well, we knew the business better than he did. For spares, breakdowns or added loads or other reasons, we'd take three anyway, and hide one in some convenient hangar."

"He never knew how many times we used three (once, even four) to serve *him*. They look exactly alike. Crazy thing about him, he would often sit in it on the ground with engines running. That can be mighty expensive. He once kept us an hour like that, with engines running, just sitting there while he played with his dog."

President Richard Nixon loved flying too, but with less pretense. He ordered copters to his retreats in California and Florida. At that time they had no adequate airplanes for long-distance transport (they use the big C5A now), so that meant ferrying the copters empty for long distances.

"Under strict squadron orders, if we removed a major

part such as a rotor blade (so the helo would fit inside a cargo plane) we'd have to fly 25 hours (with it re-installed) before we could board the President. So we flew out to California — a three-day trip."

Nixon once helicoptered into New York's Central Park. The pilots remember it as one more "hairy" experience as they didn't have the best navigation aids. But they remember Nixon, like Kennedy, as a gracious one to serve. It was so even on his final flight to Andrews Air Force Base, when he left Washington in disgrace.

One incident involved Nixon's setter dog. Handlers dragged the scared animal aboard, and it smeared dog turds all over the carpet. Marines ferried the copter back to the Sikorsky plant in Connecticut for repairs. That dog accident cost a small fortune. Afterward, they covered the floor when Nixon's setter came aboard.

Gerald Ford enjoyed the helicopters, and showed it. He was said to fly almost as much as Nixon. On one occasion, with a storm battering the White House, Ford griped loudly because he was unable to copter out to Andrews for a trip in Air Force One. To go by limousine was, for him, a hardship.

This helicopter transit line handles whatever the White House orders. That may include the Vice President (in "Marine Two"), cabinet members, high government officers, friends, and relatives, as well as visiting foreigners. On less frequent occasions, HMX totes the White House press corps in a CH53 (like the VH3) or other helicopter, but not from the White House pad, and the reporters must pay nominal costs.

Carter Cuts Back

President Jimmy Carter cut back on helicopter use, and he made a show of the frugal appearance. In May 1980, he decided to see for himself the Mount St. Helen's eruption disaster in the state of Washington. As always, the presidential decision sparked a great chain of actions.

Part of the burden for the visit fell on Charles J. Nole, pilot of the Army National Guard, Fort Lewis, Washington. He had orders to lead the President on tour of a devastated land and overcome all hazards, especially nasty weather. The state coveted a federal disaster designation, with relief funds attached.

Preparations for the visit took two full days. The White House staff moved a communications satellite in orbit. They checked hotels, airports, stores, streets, restrooms, schools, every place their Chief might visit. Then came, one by one, a string of transports — a Boeing 707 with at least fifty more White House aides, another 707 with ninety members of the White House "News Pool."

Two hours before the President was due to land, the Secret Service grounded everything within a wide radius of Portland Airport, nearest field big enough to accommodate Air Force One. A C5A (largest American aircraft) followed Carter, with three UH1N Hueys from HMX1 and their crews. They wheeled out the Hueys and unfolded rotors, ready to fly in minutes.

As Nole recounted it, "That morning it was touch and go. Scud clouds, a ceiling right on our heads and all over the mountain. At higher elevations it sat on the ground. Nobody saw the top of St. Helens that miserable day."

The flight got away under a glowering sky, steady drizzle, and sifting volcanic ash. Nole said they flew toward the west, where the weather trend was good. They made a grand tail chase — nine large copters — six in Army colors plus the three Marines.

"I heard the press grumbled about the noise. Well, the Chinooks they flew in don't have much sound proofing. And they have a bunch of chattering gear boxes inside."

Nole led the Presidential party low and slow over the Columbia River to Kelso, Washington, then north and east to the volcano's devastated side. Meanwhile the cloud ceiling cowered lower and darker, "like the inside of a bear's throat." The bad weather came largely from the volcano itself, so it grew worse as they moved in closer. At the peak itself, the clouds sat on the ground.

"I was scared that Spirit Lake would be socked-in — just what we wanted the President to see. But in time, we found a saddle under that overcast, big enough to squeeze helos through, one at a time."

"Pretty bad stuff there — messed up terrain, trees knocked down everywhere, rocks and slides, and steaming craters. Worse than any battlefield I've ever seen."

"The President's pilot stayed right on my tail. For a Marine he was good, like he was tied to us. We came to a big open bowl where the overcast was higher, with Spirit

Lake in plain sight. We did a round robin over the blasted land for all those visitors."

They found the ruin depressing. "I could feel the tension on the radio. Nobody said a word. It was only four days after the eruption — and hot, under a cold sky. Weird weather! We could feel the heat rising from the ground into clammy air from a cold front. We even hit a snow squall.

"I wonder what the President thought of that May weather. But I think he saw it all." Nole, a warrant officer third class wearing Senior Army Aviator wings, had the honor to fly the Commander-in-Chief.

The copter-borne President saw enough. He duly declared the "Disaster Area," with the money to help the injured and destitute. Nole's three-hour hop helped to bring relief and restore the worst damage from the worst eruption in American history.

Ronald Reagan's White House takes threats seriously. The crews claim they "don't see much of him" as he comes and goes. Helicopters fly for him in the usual pairs, with an escort of two tandem-rotor CH46's. The security arrangements are the most intensive to date.

The White House itself is protected like a fortress, and few other helicopters are allowed in the forbidden zone of Washington. When the American Potentate arrives at Andrews in Air Force One, the Secret Service grounds all air traffic, the airfield and ground traffic stop dead, and uniformed troops stand alert everywhere.

The scene has the marks of a real exercise in protection, as if for imminent attack. Mysterious civilians show up and eye people at the terminals and on the roads. Doors are barred. Nobody is allowed to move. At least four helicopters stand ready, with engines running or warmed from time to time.

The alert continues while the President remains on the ground, until long after his helicopters or the big Air Force One goes off into the sky. The arrival and departure exercise can take an hour or longer, for this transfer point in Presidential travel. The trip to the White House takes a mere 8 or 10 minutes.

Since 1957, seven Presidents have made full use of their chariot from the back yard of the White House and everywhere else they choose. It is royalty American style, a la helicopter.

2.

And Sudden Rescue

Those TV viewers saw what the helicopter does best.
— Don Usher, Park Police Pilot

He stood poised, his mind focused to make it as routine as possible for a first parachute jump. Andre Roux, trim and fit, age 20, waited at the open door for the word *go,* at 2,000 feet above Yverdon, Switzerland. The word came. He jumped.

Instantly he was jolted and yanked forward. Horrors! — His chute caught on the tailwheel of the plane! He was twisting, pulled tight in the windstream, tangled in the shroud lines of his chute. He extended his arms to stop spinning, but he was caught like a fish on a hook, towed by his chute harness in the frigid wind.

The pilot throttled back to 50 MPH, to ease the pressure on Roux and the airplane, and conserve fuel. He turned onto a race-track course. By radio he alerted SAR, the famous Swiss Air Rescue Service.

It took SAR an hour — the hard ones always take longer, they say — to work out a solution to this bizarre predicament. Besides the extreme danger to Roux, there was the safety of the big Turbo-Porter plane and its seven occupants. To land with the man dangling would kill him sure, and likely even cause the airplane to crash.

But SAR has a reputation for doing the impossible in the Alps. Soon, two pilots and another chutist took off, in a fast Alouette III helicopter. In minutes they climbed up over the orbiting plane, while rehearsing their plans. With flailing rotors on top and tail, a copter is hazardous to fly near the slipstream of a fixed wing, and these men had no experience of that kind anyway.

Pilot Andreas Haefele later said he just "fixed attention on that airplane's left wing tip, then we inched down as close as we dared."

Meanwhile the other pilot, Adolf Rufenacht, hooked a cable to the chute chest harness of the third man, Pierre Jomini, an experienced jumper. Now he winched Jomini down 30 feet beside Roux.

Two dangling men whipped in the maelstrom behind the copter and plane while Jomini first made sure Roux was conscious and able to handle himself. As Jomini dropped down close, the wild currents swept him away and forced him to winch back up. He tried a second time — with the same result.

On the third try, Jomini reached out and clutched Roux's shroud lines to pull himself close. He found Roux bitterly cold but alert. As Jomini related it afterward, "I roared over the extreme noise, 'I'm going to cut your shroud lines. When you fall, count to three — *three* — then open your emergency chute.'

"I used the knife. In a split second the ropes separated and Roux fell off, like a stone. A white spot grew fast down below; his parachute had opened, and I knew he was safe!"

Now free of the trailing weight, the airplane suddenly rose ahead of Pilot Haefele who slowed the helicopter while Jomini was winched back to safety. The trio had trouble believing the whole episode only took 5 minutes.

But those in the airplane thought the opposite. They came down relieved to land safely — with only 15 minutes' fuel left in the tank.

Lucky Roux found soft landing in a potato patch. The SAR copter took him to a hospital where doctors found him in good condition, aside from a little hypothermia.

That rescue occurred on September 28, 1980, a classic that showed the extent of human ingenuity, without equal on earth today, yet a real commonplace. In the time-honored way of airmen, Roux eventually made his second jump, but then for him it was routine, anti-climax.

In the first recorded copter mercy mission, October 3, 1944, Commander Frank Erickson, U. S. Coast Guard, flew a Sikorsky R4 from New York City to Sandy Hook. He took life-saving blood plasma for more than 100 sailors injured in an explosion. For over four decades since, helicopters have served above expectation as well as speed.

Ship Rescue Dramatic

Some 10,000 miles away and five days after Roux's incident, began another "commonplace" rescue of different dimensions. Fire forced 524 people to abandon ship in the storm-ridden Gulf of Alaska.

Six helicopters rushed in from distant places. Together with Coast Guard and merchant ships, they carried out one of the greatest sea rescues in modern times. Aircraft and ships of the U. S. and Canada, civilian and military crews traveled long distances in a growing storm.

Most of the 319 passengers aboard the Dutch liner Prinsendam were elderly, some in failing health, four in wheelchairs. It was midnight of October 3, zero hour of their third pleasurable day. The cruise to the Far East was into its first period in the open sea, 195 miles west of Sitka, Alaska.

Suddenly, fire broke out in the engine room. As precaution, the captain routed passengers from dance floors, lounges, and staterooms. Few wore adequate clothing, not knowing what to expect; none expected the worst in such a modern ship.

Blazing diesel frustrated the firemen. In two hours it forced Captain Cornelis Wabeke to broadcast SOS and prepare to abandon ship. His message called out rescuers in the face of an impending typhoon and black darkness, in the long nights of Alaska fall. The village of Yakutat, Alaska, 140 miles northeast was the nearest land.

Six large helicopters and two transport planes took off from bases as far as Kodiak and Anchorage, over 385 miles north; Vancouver and Comox, B.C. 700 miles to the southeast. Coast Guard and merchant vessels promptly turned about to converge on the Prinsendam. Nearest ship, the supertanker Williamsburgh reversed course at a point about 100 miles south of Prinsendam. Fortunately, the five-deck tanker had two helicopter landing pads and plenty of room for survivors.

From Kodiak the Coast Guard sent two Sikorsky HH3 helicopters plus a four-engine HC130 airplane to direct action from overhead at the scene; two more HH3's sped out from Sitka. At that hour, a U. S. Air Force HH3 began the long

Hovering USCG HH3, with rescue basket on hoist over lifeboat.

Fire aboard luxury ship marooned 524 occupants; ship later sank.

Copters lifted hundreds of survivors from lifeboats in stormy sea.

flight, 420 miles non-stop from Anchorage. It was refueled by an escorting HC130 tanker plane; it flew over 11 hours without shutting down that day. And a pair of big tandem rotor CH46's of the Canadian Armed Forces began the flight from Comox, British Columbia, 700 miles south.

When, at 4:54 a.m., Wabeke gave the dreaded order to abandon ship, passengers and crew began scrambling for the lifeboats. In the stormy darkness that took almost two hours. Not long after the boats touched water, the helicopters arrived and began lifting survivors in their rescue sling-baskets.

As the Williamsburgh hove to, the rescuers immediately saw that these survivors would never make it safely by rope ladder up the side of the big ship from lifeboats. The copters began crowding people in their small cabins, up to eighteen at a time, to land them on deck.

In one lifeboat, a disabled elderly woman was knocked unconscious by the swinging rescue basket. The copter brought her out in the same basket; she revived safe and uninjured aboard the supertanker. Nearly everyone including crew members got seasick in the tossing lifeboats. Many were in panic or shock.

Air crews said it took three or more minutes to winch each person up to the hovering copter, half an hour and longer to pack in a full load. When the Coast Guard helicopters shuttled off to Yakutat to refuel, they took survivors there. They packed in as many as weight limits permitted. The helicopters also ferried blankets and supplies, and firefighting gear.

One Coast Guard crew reported taking 110 persons from lifeboats. In all, they retrieved 359 of the 524 passengers and crew. From the Prinsendam they removed fifteen passengers and twenty-five crewmen who had stayed aboard to fight the fire. The captain left last, at 2:30.

That night the Williamsburgh, with 380 survivors, turned about to Valdez, Alaska, whence she had departed two days before. Others went to Sitka in merchant ships and Coast Guard cutters.

That was not all. In the confusion, eighteen passengers and two para-rescue men from the USAF copter were left behind, in a small lifeboat. When that was discovered at 9:16 p.m., a cutter set out in the storm, and finally found them riding 35-foot waves. The cutter retrieved the last of the survivors at 2:30 a.m. They had spent over 18 hours in the open boat.

To rescuers it was miraculous that medics found no serious injury among the 524 survivors of the prolonged ordeal.

There were minor mishaps — a broken hoist cable on the USAF HH3, and a loss of flight instruments that grounded a second Canadian CH46. The Prinsendam burned for a week; she resisted all efforts to tow her to safe harbor, and sank in 8,000 feet of water.

Fire Worst Peril

Fire is the worst of perils of the sea or air. Fire consumes the oxygen we breath, it creates searing heat, smoke, toxic gases, and disables machinery. Firefighters know it as a dragon that feeds on itself.

The hardest part of the Prinsendam rescue, the Coast Guard found, was lifting men and women from the tossing boats in seas as high as 35 feet, in darkness, rain, and winds gusting to 50 MPH. Seated up front at their controls, the pilots had to hover blindly, talked into position by crewmen hanging out of open doors so they could see the small boats directly below.

Commander Richard L. Schoel, USCG, asserted that the long time and distances, plus the bad weather hampered the work. And three of seven Coast Guard copters available were out of service.

The tanker Williamsburgh, said Schoel (who headed the rescue task), gave great advantages with her roominess and her two helipads. The other ships, the faceless crews, and yes, the passengers who found sources of courage in the dead of night, all together brought a happy ending to a near-disaster. They combined skill, courage, daring, and readiness, and not a little luck — to slay the dragon, fire at sea.

It was fire again, and again the combination of superlatives in helicopter rescue from a large Las Vegas hotel. This time two city policemen on routine patrol led it. Passing overhead in a Hughes 500 copter, they became curious about a trash fire at 7:15 on a November morning.

Before their eyes, a geyser of smoke engulfed most of the tall, bulky MGM Grand Hotel and Casino. The fire raged out of control on the first two floors, but smoke and gases swirled far around its twenty-six stories.

Circling the 2,076-room hotel, the pilot, Sgt. Harry Christopher, saw people gathering on the roof of one wing. He decided to take them off to safety, but when he landed on the roof they all rushed over in panic.

The other officer, Tom Mildren, got out to calm them while the pilot shuttled — four or five at a time — down to the parking lot. With a fire and a hotel so large, Christopher saw one helicopter could never handle the load, so he radioed for help.

The crowd grew on the roof. People broke windows and screamed on balconies. As more helicopters responded, from the local hospital and commercial owners, they set up a shuttle from the hotel roof to the parking lot. Christopher called in a second Las Vegas police copter, a small Hughes 300C, to station low overhead for air traffic control.

The sergeant admitted overloading his 500C — sometimes even six passengers in the four seats — when he flew off the roof into virtual autorotation landings[1]. Often the landing gear skidded 15-20 feet as he touched down.

Nine Air Force rescue copters arrived from local bases; they took victims off balconies with the aid of rescue hoists. While in hover halfway over the building eaves, crewmen tossed their rescue hooks onto balconies below, but the pilots could not see below the overhanging roof ledge. Police and civilian copters came to their aid in this tricky business by moving in close, and directing pilots by radio.

Altogether, thirty helicopters — private, police, military — from as far away as Rialto, California, responded to the emergency. They put seventy-five firemen on the roof to go down inside. They carried engineers up to open ventilators high on the building. They rushed spare oxygen supplies, for firefighters, from 80 miles away.

Help From Far, Near

The rest of this major fire story involved firefighters from far and near, police, medical, and rescue personnel. Not knowing the extent of the fire, pilots had to assume the roof would burn out or collapse. They kept alert to get away at any moment.

Sgt. Christopher noted there was no turbulence and no

[1] Full explanation in appendix.

adverse wind. But pilots contended with other troubles – confusion in communication among the many aircraft; traffic control; smoke and gases from burning plastics were toxic to breathe, and could kill their engines.

"Some of the Air Force helicopters had no VHF radio channels so we could not talk to them, and they had no link with firemen. When they dropped firefighters on the roof to open doors, smoke and flame gushed out and their downwash blew it on the firemen – who had to run for their lives.

"We had to land too close to the source of fire. We lifted guests who didn't understand English, and they were in panic. We had to land with tail rotor over the edge of the roof so they would not run into that. We had loose boards and items of clothing that could get into the rotors, and the big sign on the Hilton hotel next door kept blinking on and off and blinding us!"

Somehow it all came off without mishap. Sgt. Christopher had a regret: "Not one helicopter took one victim to the hospital." They landed every one of them in the parking lot for first aid.

Eighty-five persons died in that fire, hundreds were injured. Most of them, in top floors, were overcome by smoke. The worst hazard – toxic fumes from burning plastics and other materials – fed into the top floors through the hotel ventilating system.

In an unforeseen emergency that came with no preparation of any kind, the faceless rescuers proved equal to the challenge. In little more than five hours it was under control. The safety of 8,000 guests, employes, and visitors in that hotel was assured while firemen quelled the blaze. At dusk the last aircraft departed as Christopher and Mildren made one final circuit.

The MGM Grand was neither the first nor largest high-rise fire rescue by helicopter. Nor last, as fire remains the ancient menace everywhere. It wasn't the last casino fire in Las Vegas either. In February 1981, while cleanup work was still going on at the Grand, fire broke out in the thirty-story Hilton. Again came the copters. Later, still another casino blazed up suddenly, this time in Reno – with still another cast of helicopter rescuers.

As long ago as 1966, helicopters snatched sixty persons from a burning building in Israel; eight more from a tall building in New Orleans, Louisiana, and 200 from a skyscraper in Bogota, Colombia. Two more occurred in Sao Paulo, Brazil, where they lifted 350 from a thirty-one-story roof; and twelve more in Kuala Lumpur, Malaysia.

In that last case, the big Bell 212 could not land on the roof — a flagpole stood in the way. Another helicopter bent the pole flat. Fire and explosion in the thirteen-story building got so intense that, minutes after the rescue, the roof went up in flames.

After a second tragic fire disaster, Sao Paulo mandated helipads on all new buildings of twelve stories or higher. In the USA, certain large cities began requiring pads on new structures, but they have not always enforced them. The reaction has not been unanimous, even where tall buildings hold thousands of people. New York's World Trade Center, for example, has 125,000 in its twin towers. But some city codes require roofs prepared to make rescue easier, or at least possible in case of fire disaster. The helicopter is the *only means of reaching these heights,* and often the most acceptable.

There is opposition. Professional firefighters highly praise helicopter rescue but disagree over such laws. They protest against enticing people onto the roofs of buildings, where the fire goes anyway. They say it is wrong to urge people onto the roof when they could better be heading for ground exits. But helicopters get there first!

Prepared or trained or not, rescuers seem to show up everywhere, as needed. Firefighters, policemen, Coast Guardsmen, and military units constantly prepare and train for emergencies. But who can anticipate every possibility?

The Las Vegas police had no training or plans for the MGM Grand. Nor did the other private, civilian pilots from far and near. Mel Larson, who flew a Jet Ranger, is himself vice president of the Circus Circus, another large Las Vegas casino — with a helipad on the roof.

Training, preparation, and suitable helicopters as well as proper gear are, of course, essential, as proved with the Prinsendam. So it always proves when police, military, hospital or fire units go into action. This challenge — to which pilots have responded since the earliest times — reaches out to the casual pilot in a trainer or crop sprayer, and he does whatever he can on sudden call. The possible often works well enough.

North Sea Rescue

An intrepid bunch of British pilots bought such challenge in a hairy sea rescue. No fire in this one, on December 15, 1979, but the North Sea was at it wintriest — storm winds up to 70 MPH with gusts over 90, 40-55 foot waves, overcast 500 feet, visibility 2 miles, blowing snow, temperature at freezing.

Bristow Helicopters' task was to evacuate 527 workmen from a derrick barge located 115 miles off Aberdeen, Scotland. Bristow normally flew to the barge on contract with the Texaco owners.

Chief Pilot David Smith got his orders at noon, after he had cancelled all flights for bad weather and hangared the copters.

The 100,000-ton barge, devised for oil platform construction and worker living quarters, had broken its 3.5-inch steel mooring cables in the angry seas. It was drifting toward the coast.

Captain Smith got off the first helicopter, an Aerospatiale Puma with seats for nineteen, in 20 minutes. Then followed twelve Sikorsky S61's seating twenty-three, manned as fast as crews could return for duty. In all, twenty-eight pilots flew the dismal 2-hour round trips to the barge.

Later, Smith wrote, "The barge helideck was pitching 20 feet in a 40-foot swell and rolling up to 4 degrees. As both crane jibs were stowed in cradles each side of the helideck, pilots were committed to reversing their aircraft onto the barge in winds gusting over 80 knots, before landing."

In other words, on a deck tossing and turning dangerously over the accepted limits for putting down these helicopters, they came in to land *backwards*!

It was orderly; passengers were ready to board so that copters stayed on deck five minutes or less, with rotors turning up. The mass evacuation came off with precision — twenty-five round trips in seven hours — four of those trips in stormy darkness.

Captain Richard Collishaw admitted he was "very new to the game up there, and cautious. I don't see much of a story in it, do you?" His question underscored the very routine-ness of this action. The pros carried on with the day's work.

Rescue efforts often turn out like that, sudden, random, counting on pilots, mechanics, and airmen alike to know, by instinct, how far they can extend the limits of skill and aircraft — *if only because failure is worse than not making the try.*

It was something like that with the National Park Service Police when an Air Florida Boeing 737 smashed the Potomac River bridge in Washington, DC. Chief Pilot Don Usher and Observer Gene Windsor grabbed available rescue gear and rushed in their Long Ranger to the scene in five minutes.

The story needs no repeating. An incredulous world saw it on television: the silver and blue copter under a 300-foot ceiling in blowing snow: plucking people from a hole in the iced-over river: literally dragging five survivors to shore. With what they had at hand, they took off in the atrocious weather, intending to look for the crash. If conditions got too bad they could return to their warm hangar.

"It only took a couple of minutes to reach the crash site [from base 3 miles away]," Usher said, "but when we got there we couldn't even see the other end of the bridge. Visibility was that bad. No one knew exactly where the aircraft was."

"Gene could see the wreckage on the bridge, however, and when we got right down on top of it we could see all the broken ice and the pieces of fuselage that the victims were holding onto."

Suddenly they brought salvation for the survivors of that awful airplane crash and the cars it crushed on the busy bridge, near the Pentagon. Although TV did not show it, Usher had full radio contact with other forces, busy as he was.

In a matter of minutes, Army and Maryland State Police copters landed on the narrow bridge span, in sequence. They carried forty-five injured survivors to hospitals. It begs nothing of the heroism of these rescuers to ask how many more might have survived if . . . .

How close even those survivors came to be left without this aid. James Watt, then Secretary of the Interior, had moved to abolish this police helicopter unit, to save money. And how close it remains today; Washingtonians complain that nothing has been done to curb the dangerous conditions at this busy airline terminal in mid-city.

The same city turns a cold eye on copters of its Metropolitan Police, who have no such rescue capability. Nor the Coast Guard with its headquarters but no convenient base there. Nor any aid from the President's squadron of

U. S. Park Police Pilot Don Usher, left and Observer Gene Windsor, heroes of Air Florida crash rescue in Washington D.C.

Marine copters at nearby Anacostia. Nor any other pertinent agency in the central city of a government colossus — the richest in the world.

But, again, in Washington the rescuers were ready, to do what they could. They put the injured on shore, in reach of medical aid. If, as in all rescues, time is urgent, the injured double the trouble. Injury makes for a different kind of rescue, where the helicopter hastens medical aid to the sick, to those who are hurt and dying.

As with Andre Roux, with the people on the Prinsendam, with uncounted other hapless ones in storms and disasters, the helicopter mocks the impossible, to make the splendid rescue a commonplace. Once more, lift is where you find it.

Today, well over one million persons owe their lives to the helicopter, this machine of all scenes, say those who like to count. But rescue, the myriad other services that helicopters perform in time of dire need, don't lend themselves to accounting. Probably it doesn't matter. The survivors of fires and accidents know rescue as a personal experience, and that's quite enough.

Police Pilot Don Usher, on receiving an award for heroism on the 14th Street Bridge, said simply, "All those TV viewers saw what the helicopter does best." But we know helicopters cannot do anything by themselves.

3.

Trauma: When Minutes Count

Half of trauma deaths occur in rural areas. Of those who reach hospitals, perhaps fifty per cent die within four hours.
— Dr. Howard Champion, Washington D.C.

They call it "Life Flight," "Air Care," "Bandage 2," or other fanciful name. It comes rushing in from a hospital — emergency nurse, medical technician, sometimes a doctor, all helicopter-borne.

It lands, colorful, noisy, as showy as an ambulance, even without flashing lights. The sick and injured hardly notice, they rather welcome this Emergency Medical Service (EMS), as it brings them critical care. This helicopter, doing no ordinary kind of rescue, comes for one prime reason, trauma.

Trauma means real hurting people. . . victims of automobile carnage, the grossly burned, maimed children, ailing babies, victims of heart attack and stroke, of stabbings, shootings, and drunken mayhem, workers and bystanders caught in the bizarre and "freak" mishaps that daily find a place to happen.

What is *trauma*? The word lends to shades of meaning: "the insult to the body"; the physical-psychological *reaction* to injury or sudden illness such as poisoning or heart attack. Without prompt treatment — in minutes — it brings shock and irreversible effects, even death.

For the San Diego County Health Department, Gail Cooper defines a trauma victim as one who is unconscious at least 5 minutes, has at least thirty per cent severe burns (half that amount for anyone under age 16), and-or unstable blood pressure, pulse and other vital signs.

The EMS helicopter evolved with the endless sources

of trauma. It is, first, a machine with equipment costing upwards of $300,000, a skilled pilot, mechanic, and medical crew. It works under close controls, with elaborate communications. It teams with fire, police and other public services, and ground ambulances. Most EMS copters serve the ill and injured exclusively, leaving to others the task of rescuing the endangered who are not hurt.

"We don't use the term 'rescue.' It's a difference of motivation. We have aircraft ready, 24 hours a day, to export the medical team from the hospital out where the injured patient is." (Dr. Dan Reich, St. Anthony's Hospital, Denver.)

"Experience shows that success depends more on the quality of the first aid rendered at the scene than on the speed at which we carry the patient to the hospital." (Dr. Gerhard Kugler, Auto Club of West Germany.)

The earnest medical leaders know the worst; if they disagree on how to work EMS, they unite on the urgency of trauma. Dr. R. Adams Cowley, Baltimore, labels it "the golden hour." Meaning medical intervention within 60 minutes of injury, and copters or other means to speed that treatment.

"The problem with trauma," says Dr. Howard Champion, Washington (DC) Hospital Center, "its death rate is increasing. It's the most expensive disease of modern society." He called it America's worst epidemic.

"Most trauma comes from auto accidents and knives and guns. Thirty-five per cent of those patients die, but we can cut that to fifteen per cent. It costs the USA well over $60 billion a year."

"Half of trauma deaths," Dr. Champion went on, "occur in rural areas. Of those who reach hospitals, perhaps fifty per cent die within four hours. Half of those could benefit from trauma therapy."

Dr. Champion once treated a woman stabbed in the neck. The sight of it so shocked a cab driver at the scene that he had a heart attack. In another case, before they could load an injured auto driver in the helicopter, chain saws were called in to cut off a long board that had impaled him.

Value of Life?

Both helicopters and trauma treatment are expensive. So how much is a life worth, in dollars and cents? Does

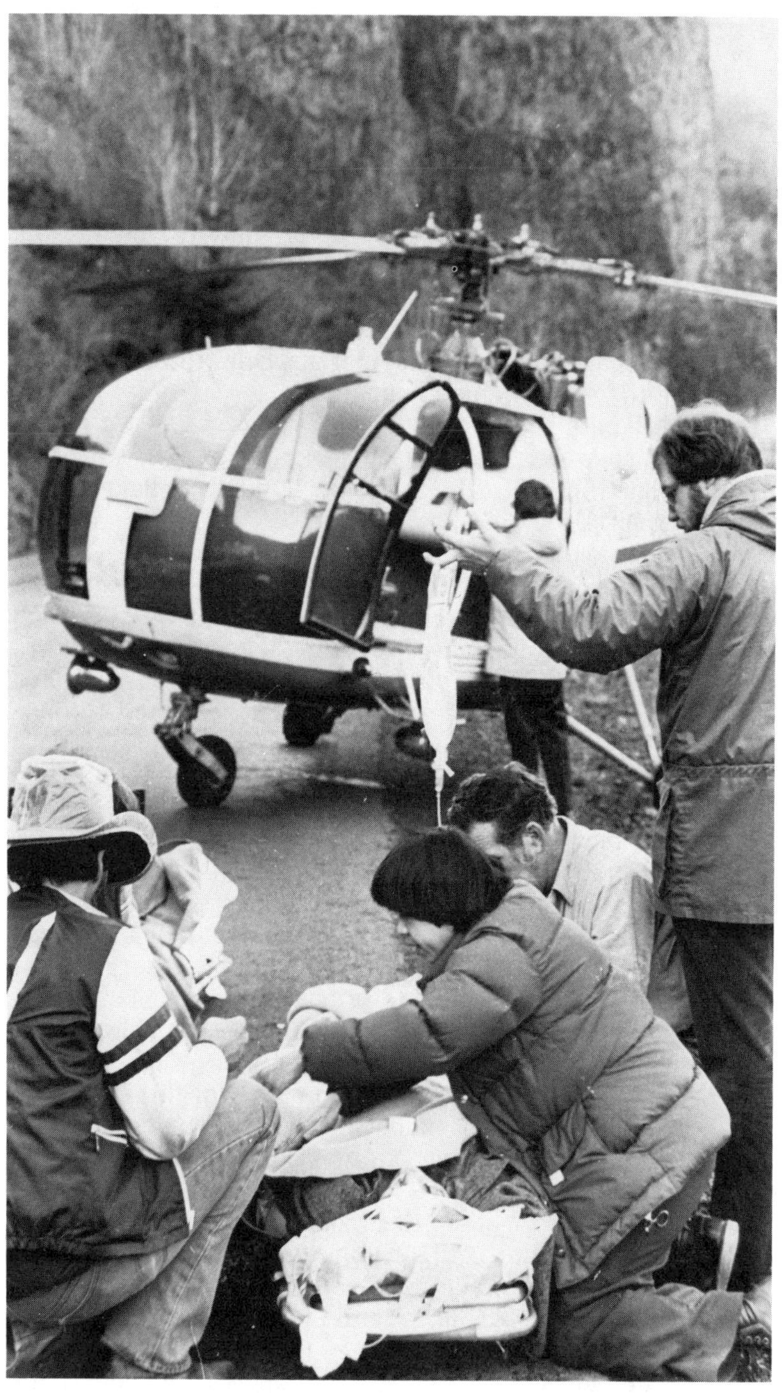

EMS copters fly direct from accident scene to hospital emergency care. This is popular type, Alouette III.

society owe everyone as much as each of us expects for ourselves, in relief of suffering? How to overcome apathy and misunderstanding of medical crises?

How much should we spend on victims of their own foolishness, of drunk driving, daredevils, "chicken" and drag racers? How far must we go to nurse shattered bodies to a semi-vegetable existence?

Such questions have nagged the medical profession for generations. With ever more medical marvels, e.g., the artificial heart and routine organ transplants, the old methods and practices are not good enough.

EMS pilot Marc Johnston, a Viet Nam combat veteran, gave an answer. "We spent an average of $40,000 to kill each Viet Cong. So why not a few hundred dollars to save good Americans?"

As we know from Viet Nam, armed forces go far to retrieve the wounded from battle. Helicopters snatched them from amidst enemy fire; jets took some of them back home in hours, to the very best of medical care.

In World War II, about ten per cent of casualties died of wounds; in Viet Nam, one per cent. Injured soldiers waited 8 hours, on average, before treatment, in World War II, but only 3 hours in Korea, under 2 hours in Viet Nam. Statistics reveal not a thing about suffering. Those were real wounded men who got the relief, the benefits of treatment, so much sooner.

The EMS copter performs three main rush tasks, delivery to trauma centers; delivery of critical patients between hospitals; and delivery of hearts and other organs for transplant, special equiment, serums, and medicine. (A few also take on rescue tasks when no other is available.) They are ready on call, airborne within 5 minutes or less, 24 hours a day, in all but the worst of icing weather or fog.

In addition, the hospital will have close communications with the medical specialists in flight; they will treat victims at the scene, on the return flight, and in the trauma center.

EMS links together the highly skilled pilot with the medical technician or flight nurse or doctor. All of them take specialized training for this flight duty. Bill Caraway, Jackson, Mississippi, pilot for the University of Mississippi Medical Center, writes:

"You never know when a request will come. . . the middle of lunch, or 3 a.m. When it does, you have minutes to fix the location, get navigation data, check weather, then

"Life Star" EMS crew of Bell LongRanger, University of Mississippi Medical Center: Pilot Bill Caraway, at left, and Flight Nurses Mary Kay Holliday, Janice Mooney, Scott Berry.

Landing pad and ramp lead into hospital's trauma center.

From MBB-Kawasaki BK117 copter on helipad, trauma patients go direct to treatment at Washington (D.C.) Hospital Center.

either take off or cancel out. There's a chain of rescue, medical, and police waiting to hear when you will arrive, and people's lives hang in the balance.

"Weather could be marginal. It could be a bad accident, at night, and the helicopter is the victims' only chance for survival. What do you do? How do you decide?

"The answers are unique to each situation, and will depend on the pilot. A veteran will make a very different decision from a new pilot who is still learning the area, deciding on the spot." His first concern: be sure he can deliver, "to go and return — so the hospital can take you for granted. You develop a sense of it. You know what to expect at a strange landing zone, to deal with bad weather, and the special problems of night flying."

Caraway began hospital work after 3,000 flying hours. He had varied experience, much night-flying-frost-control in California. He learned to deal with all sorts of conditions and hazards.

Low hanging wires are among the worst dangers, Caraway writes. "You have to fly low and close to the ground and you get wary. You fly over power poles. Wires hang below them. Fortunately, floodlights on the helicopter make wires stand out against a dark background."

After six years and 5,000 hours in this work he is satisfied to "know I'm there for the benefit of the flight nurse. . . She (or he) is familiar with me, and can assert authority over law men, fire people, ambulance drivers, others who might get in the way."

Flight Nurse Pat Roberg said her first duty is to "get in touch with the victim, let him or her know help is there, and make ready to fly to the hospital." The nurse begins the course of treatment for trauma. This helicopter carries an array of apparatus, oxygen, blood plasma, defibrillation and other devices. There are special kits for heart cases, burns and infants.

Tragedy Recounted

One who knows this humanitarian service, Mrs. Ronny Herring, Redland, Oregon, relates it in the worst of personal tragedy. "People who haven't experienced such an emergency just do not know."

An auto accident killed two of her three children. A son, 13, died at the scene; another son and a young daughter critically burned in gasoline fire. She herself had severe burns. She was relieved to hear the helicopter arrive at the accident scene. "I know those wonderful rescue people did everything they could."

"I was conscious through it all. At first I didn't know the helicopter took Sean and Shannon to the hospital." Not until later, after treating her trauma, did they tell her Shannon was in critical condition.

It happened on the edge of town, 12 miles from the hospital. Life Flight got the call at 6:22 p.m., lifted off at 6:24, landed on the freeway ramp at 6:31, loaded patients and took off again at 6:45. It was back on hospital pad at 6:50 — 28 minutes after first call, Sean and Shannon reached the city's best burn center, into full treatment for trauma.

Mrs. Herring came in by ground ambulance. She recovered, but not her daughter, who was only 8. They were about to amputate her leg when Shannon died, four weeks later.

Sean, 11, had second degree burns on much of his body, including his face, and threat of disfigurement, but weeks in the burn center and treatment in the following months brought him to full recovery.

Does he remember that night ride? "No, I was out of it, for a whole week." Now he nurtures ambition to fly helicopters himself.

Insurance paid for the helicopter but the Herrings themselves would have paid, if necessary, for their surviving child. "For all we know, it might have made the difference that saved Sean."

Dr. Champion cites the millions of civilians injured each year, 10 million disabled. Accident is the fourth leading cause of death — the leading cause among Americans under age 38. "Twenty per cent of trauma victims die at the scene; eighty per cent would survive if properly treated."

EMS saved the life of Mike Walther, 17, of Avon Lake, a suburb west of Cleveland, Ohio, who took a bullet in the chest while on a hunting trip in the southern end of the state. It happened when another man in the party hit a deer in the forehead, while Mike was only 38 feet away.

The slug (a "deer load" from a 16-gauge shotgun) glanced off the base of the antlers and struck Mike and lodged one inch from his heart. His companions dragged him out of a ravine while others went for help. Paramedics gave first aid and carried him to a nearby road. They summoned Life Flight of Grant Hospital, in Columbus, which is about 100 miles away.

"Without that flight," Mike says now, "I probably would not have made it. I was awake the entire time; they had everything inside the helicopter. The female doctor on board [actually Flight Nurse Cindy Crable] put me on a life support machine and inserted a tube in my collapsed lung. I was in the hospital in Columbus in minutes.

"That was hours before my brother (who was in the hunting party) arrived by car, partly over rough rural roads that would have been dangerous and uncomfortable."

Dr. James E. Nappi, assistant professor of surgery at Ohio State University, and an associate at Grant, commented on the urgency of treatment. "Even with Life Flight it can take two hours to move the patient from a rural area to a metropolitan medical center. Another half hour may be spent in the emergency room. If two to four hours are needed to transport the patient by road there may not be enough time remaining for the surgical team to do the necessary vascular repair."

The copter does smooth the ride to the hospital, in the experience of Fireman Terry Bowman. He was caught off-guard in a flash fire and suffered second- and third-degree burns. He could not bear to be touched, nor lie on a stretcher. He walked rather than ride an ambulance a few yards to the waiting helicopter.

"I could feel every jiggle. Believe me it hurt bad! It was a relief to know the copter was coming. It helped when the flight nurse wrapped me in a sterile sheet. When we took off it surprised me how smooth it was!"

"When I turned my head I could see the whole city. Then we headed down to the hospital. Only took 8 minutes — a great ride."

Bowman believes he owes his life to the helicopter and the burn center. He remained conscious throughout, but felt himself sinking into shock. He remained critical for nineteen days, and in the hospital another five weeks.

Originated in 1972

The concept of the dedicated EMS helicopter originated in 1972. The Loma Linda University Medical Center, east of

Los Angeles, bought two roomy Sikorsky S58's and used the services of an operating company to maintain and fly them. About ten years later, the hospital dropped its EMS for "economic reasons," but it continues receiving trauma patients through police, private, and other helicopters.

In 1972, St. Anthony's Hospital, Denver, began flying an Aerospatiale Alouette III off its roof landing pad. Airwest Helicopters, Fort Collins, Colorado, provided the copter, the pilots, and maintenance services. Airwest specialized in flying the high altitudes and hot thin air of the Rocky Mountains.

Since then, St. Anthony expanded to serve thousands of clients. The hospital added fixed-wings for critical patients going or coming over 200 miles, even far out of state.

Depending on how they are defined, hundreds of helicopters fly for hospitals today in the United States, more in Canada and other countries. And most modern hospitals feature the roof or parking lot landing pad. Hospitals directly operate about 100 EMS copters, and their numbers continue growing.

What began with single-engine models such as the Bell Jet Ranger and the roomy, high-lift Alouette III has come to favor twin-engine versions, e.g., MBB BO105, Agusta 109A, Aerospatiale TwinStar, Bell 222, and the large Boelkow-Kawasaki 117. These cost $600,000 to $1.5 million, plus communications and other equipment.

The largest EMS program, at Hermann Hospital and University of Texas Medical Center, Houston, operates five TwinStars and two turboprop airplanes. It carries more than 3,000 patients a year.

EMS grew readily in the West, with its mountain and water barriers, sparse population, and open spaces, and has recently spread to most of the fifty states.

In Arizona and Illinois, state police fly "medevac" duty but Phoenix and Chicago hospitals run their own helicopters. In Phoenix, three hospitals operate helicopters in competition, and carry more than a thousand patients a year. In Salt Lake City, the Mormon hospital operates one helicopter while the University of Utah Medical School and three hospitals share another. And ten hospitals south of San Francisco work together with one helicopter, while Stanford University operates another in the same area.

The emergency helicopter, believes Barbara Phillips, of

Vienna, West Virginia, saved the life of her son, Christopher, 12. In a 40-foot fall onto rocks at Ash Cave, in southern Ohio, he sustained a fractured skull, crushed chest and lungs, fourteen broken ribs, and multiple fractures of one arm.

After park rangers and paramedics took him to a nearby hospital, Life Flight Pilot Murray Price and Flight Nurse Joyce Zwart picked him up and flew him to Grant Hospital, Columbus.

Young Chris, Mrs. Phillips pointed out, was in grave danger. "He was lost and revived three times — first in the park, again on the Life Flight, and once again in the hospital. His heart and respiration had stopped entirely."

The same helicopter saved Charles LaBarge Jr., and his son, Christopher, 26, of St. Louis, following an airplane crash in a high density area of Columbus's east side. En route back to St. Louis, the senior LaBarge crash-landed the single-engine Piper Lance on July 8, 1983, when the engine failed at 1,400 feet altitude.

In the emergency, LaBarge headed over a tree and under a power line toward an open field, then struck an unseen telephone pole, which ripped off one wing back to the gas tank. Fire threatened the two men so they scrambled away from the wreckage despite serious injuries.

With an injured spine, LaBarge said "it felt like my ass was on fire. I don't know what caused it." When he overheard the paramedics mention Life Flight he cried, "Get the helicopter. Get the thing."

LaBarge learned to fly at age 51 and got a helicopter rating at 58, so he knew what to expect. "My intense pain was not helped by the brief period of vibration in translational lift (as the copter accelerates in forward flight). I knew what it was and expected it."

"I also knew that it would have been much worse by ground transportation. I wondered how it might seem to an experienced patient, and I asked the pilot about that. He said he knew he didn't have to tell me, but they do tell the inexperienced patients."

Grant Life Flight uses a well-equipped new MBB BO105 in what hospital spokesmen call a community service. They stress that only the "patient's condition, medical services needed, and the medical facility best able to provide the service" dictate where the copter takes the patient. This EMS service is not self supporting.

The hospital contracts with Omniflight Helicopters, Janesville, Wisconsin for the helicopter, a backup, three pilots, and a fulltime mechanic. The hospital pays $27,000 a month, plus running costs for fuel and oil, and $2.75 an air mile.

Grant attempts to respond to all calls, and hospital employes say they bill patients "as seems appropriate." The rate is $20 for liftoff and $9 per one-way mile. It makes no charge if the helicopter flies but is not needed.

Maryland System Largest

Maryland has perhaps the largest state-operated system, with a fleet of ten Bell Jet Rangers and twenty-eight pilot-paramedics of the state police. From five scattered locations, each within minutes of a trauma center, they cover the entire state. Since 1970 the state police have aided more than 20,000 persons in rescues.

"We destroyed the myth," said their commander, Captain Gary Moore, "that helicopters could only be used in rural areas. In urban places we overfly congested traffic. And that will only get worse."

"Medevac," including rescue of the uninjured, takes thirty-eight per cent of the Maryland police flying hours. The rest goes for law enforcement, surveys, and photography.

This agency succeeded through a real accident. When a police Jet Ranger rushed a close friend of then-Governor Marvin Mandel to the hospital from auto accident, it saved the man's life. Mandel applied his political talents to make Maryland's EMS permanent.

The police work closely with Dr. R. A. Cowley and an institute he founded in emergency medicine at the state university medical school. The doctor, already widely known as a heart surgeon, began pioneering in trauma medicine.

In another EMS variation, Acadian Ambulance Service, Lafayette, Louisiana, flies Bell Long Rangers for thousands of workers on oil platforms in the Gulf of Mexico. With a fleet of ground ambulances and fixed wings, Acadian serves forty hospitals in about a third of the state in addition to the offshore side. That puts the majority of its clients within 75 minutes of a hospital.

Similiarly, the Los Angeles County Fire Department, responsible for the frequent threats of brush fires, does rescue and EMS work with Bell 205's. It has hoists, slings and other equipment, and flies paramedics throughout a large territory, 4,070 square miles that include Santa Catalina Island, and Lancaster in the Mojave desert.

Controversy simmers over public service-operated vs. hospital-operated EMS. Medical specialists, dedicated to saving lives and suffering, question how the police, trained and ready to subdue or shoot the wayward, can turn life-saving talents on and off at will. Moreover, law enforcement inevitably will conflict with EMS calls.

On the other hand, hospital copters sit idle long hours waiting for patients. Like police and firemen they spend time being ready. Helicopters and EMS crews are expensive, so the hospitals subsidize the cost by charging higher fees for other patients. And rival hospitals claim that most patients go to the helicopter-base hospital in any case.

However it is done, it seems clear that we need both hospital- and public service-operated copters, for medical and rescue needs.

In Canada, with its long distances and sparse population, all provinces west of Quebec have helicopter and fixed wing emergency airlift. The Provincial Health Programs contract with commercial operators for helicopter services.

In British Columbia — 366,255 square miles, larger than Texas — the program charges a basic $28 for the first 40 kilometers, plus 29 cents an air-km., to a maximum of $180, even if it uses more than one aircraft. Non-residents pay $72 plus $1.37 a km (no limit), but no charge for medical supplies, including oxygen.

The system is there and is readily used, Canadians say. In BC no hospitals have their own helicopters, but the Emergency Health Service will pay for transport of patients anywhere within the borders of the province. In Alberta, a similar health system will order transport of patients from outside its borders, even outside the country, if necessary.

In Ontario, the government has contract EMS helicopters, with full equipment, at five locations from Toronto in the southeast to Sioux Lookout, in the northwest. They fly the Bell 212 and the Sikorsky S76, with critical care attendants in the crews.

The Swiss Air Rescue Service, which operates 90 per cent on donations and public subscriptions, blankets its mountain country with helicopters — none are more than 15 minutes from any location. SAR frequently sends doctors along to stabilize critical patients at the accident scene.

Began With Auto Club

In West Germany, in 1970 the Automobile Club began rescue service with its first helicopter. All helicopters are integrated in the emergency system as an extended arm of a hospital, according to Dr. Gerhard Kugler, who directs the service.

Today, he said, "there is a dense network of twenty-nine rescue stations covering nearly 80 per cent of (the country). . . . They are placed at well-equipped clinics and are on standby daily from 7 a.m. to sunset for emergency purposes." The radius of action is limited to 50 km, about 35 miles.

"The helicopter as the most expensive means of rescue should bring the most qualified aid. . . . Each helicopter carries an emergency doctor and a medical attendant . . . . We station helicopters at hospitals to get the doctor aboard as easily and quickly as possible — with takeoff in 1 or 2 minutes. . . ."

In the USA, civilian EMS grew out of the so-called "medevac" lift in Korea and Viet Nam. Medical and emergency experts interested in trauma agitated for helicopter assistance. After a trial with four units in 1970, military copter rescue units inside the country began offering the service in MAST (Military Assistance for Safety and Traffic). MAST is managed by the Coast Guard.

The Defense Department allocated some thirty squadrons to MAST, most of them in the South and West. The majority, Army National Guard or Army Reserve units, operate from military bases with Bell UH1 Hueys and OH58's, and fly for the training experience.

They make no charge for service, but operate under restrictions as to hours and missions, and don't always provide "instant" service. Their rules forbid them to compete with civilian operations. They have carried out many thousands of missions in EMS.

Certain communities found this service attractive as "it costs us nothing." (They pay no heed to the high military costs paid by tax funds.) And hospitals, which operate as a business, found EMS useful to help fill their beds.

Dr. William Smith, who heads Houston's Hermann

Hospital, admits that Life Flight loses about $1 million a year, but the increase of patients and their longer stays more than make up for it.

Doctors agree that EMS works best in conjunction with a trauma center that is staffed and equipped fulltime by specialists in trauma medicine. It does far more than the nominal Emergency Department, and usually operates separately.

Hospitals lease helicopters at $25,000-$40,000, per month. Add to that the cost of staff, personnel, equipment, and supplies, and the monthly costs nearly double. Such expenses strongly influence the decision whether or not to institute this weapon against trauma.

The fees they charge patients — ranging from $150 per flight and $3 a mile up — don't cover the costs, so the hospitals subsidize EMS. They have difficulty justifying such costs, which are not all covered by patients' health insurance either.

"We can't prove it is cost effective," said Dr. Boyd Bigelow, after ten years of running a trauma center at St. Anthony's, Denver. He told a seminar that, "we know it is important, the record is clear, individual cases are dramatic, but we cannot prove it in dollars and cents for the auditors."

Trouble is, he said, defining trauma, assessing results with and without the special treatments, are all very difficult to measure.

Lucky President

Still, the results seem clear. When President Reagan took an assassin's bullet in the chest in 1981, he was lucky. Only months before, the District of Columbia had designated four trauma centers, among them George Washington University Hospital, where Reagan was rushed, through city traffic, from about one mile away.

Two other victims went there at the same time, and a fourth, a DC police officer, went into trauma treatment at Washington Hospital Center. All four survived.

The doctor who headed the GWU trauma team credited the "systems approach" for their good recovery. Dr Joseph M. Giordano wrote in the Washington Post that on arrival, Reagan had lost thirty per cent of his blood, and was near a state of shock, "which might have become irreversible. He would have 'crashed.' "

"He was in resuscitative area," Dr. Giordano added, "for 35 minutes; he came in with a blood pressure of 80 and left with a blood pressure of 160. More good work was done in operative and post-operative periods, but . . . his vital signs were stablized."

Dr. Giordano used the incident to showcase the need for trauma centers. He cited the great gains in battle casualties, and asked, "has this country organized the existing technology to provide this kind of critical care?"

No helicopter figured in Reagan's bout with trauma, but what if the attack had occurred somewhere else, say 75 miles from a trauma hospital?

The problems are there, accidents never go away, and the cost of treatment is very high. But trauma remains too expensive for all. The earnest administrators, nurses, technicians, doctors, surgeons, and specialists talk endlessly about technology, statistics, and the rest of it. They don't often mention the humanitarian side, that of suffering, the very motive that brought them into medicine, though that is ever in their consciousness.

There is still another side, illustrated in a story told by a flight nurse. A battered victim of a tavern brawl was rushed to the hospital emergency room with serious injuries and low vital signs. The medical team was ready.

As they wheeled him into the trauma center and revived him, he peered at the big medical staff, the glittering devices and equipment, the bright lights. With wide eyes he said, "Hell, I can't afford this. Take me to the VA hospital."

4.

The Great Bug War

The textbook on eradication is fuzzy; each insect in a new environment is, by its nature, an unstudied phenomenon.

— Jerry Scribner, Director, Medfly Project

The whop whop of rotors breaks the midnight stillness. Blue flame exhausts from whining turbine engines. In the close-guarded compound, mechanics fuss with six loaded Hueys.

They zoom upward in a tail-chase crowding together over the trees; they shift into a wide formation and take a bee-line course at 120 knots toward the horizon — the drop zone.

As they pass a bright searchlight the lead pilot barks "Now." They head into the run at 70 knots. Relax! crews tell themselves, only 14 damned miles of this, then turn back. And back again. Forty miles up the way, another six pilots fly another zone the same way.

From each dim-lighted cockpit eyes scan the sky and ground for hazards, for other aircraft, towers, wires, fog, signs of hostile fire, birds. At the end of each straight run they reverse and shift course over an adjacent strip.

In two hours they'll go to base for fast reload with engines running, and come back. Return again and again: six hours of night flying over blocks of enemy land, never more than 300 feet above the rough terrain.

Pilots are well trained and motivated, but none anticipates that this monotony will go on for months. Later, the lead pilot, Darrell Ward will frown, "We thought we'd have an easy go of it. Only, it kept up for dreary months."

The worst of this war is not the enemy's firepower but his stealth. This is not jungle but a metropolis, the urban rim of the south San Francisco Bay. The front is the houses, farms, shopping centers, factories, shops, and orchards of "Silicon Valley," and beyond.

These are not soldiers but itinerant pilots on a novel commercial job. The enemy is bugs, tiny Mediterranean fruit flies. And it is a skirmish in a hot political jungle.

The bug is not the innocuous house fly. The war began with the finding of one tiny fruit fly larva in the city of Santa Clara – it could bring worms in apricots, cucumbers, avocados, oranges, apples. The maggots send fruit rotting to the ground, then the fly emerges, days later, to multiply its kind, over and over, in an exploding cycle.

Indeed, it is a war on maggots.

The helicopters carry no armaments and make no more than engine and rotor noises, but it is racket enough for sleeping residents. No body count here, though sprayers have flown from a civilian cemetery. Gun threats come not from enemy troops, but irate citizens.

The copters are not Army Hueys but lookalike, sound-alike Bell 205, 212 and 204 models, built or modified for civil use. On parallel paths, six copters atomize malathion-poisoned fly bait. They cover a 1,200-ft. swath, nearly a quarter-mile wide.

This war posed a question about helicopters in certain minds, folks who say "everybody knows" they crash and kill, and fly too low, raise dust and noise, violate privacy, and scare animals. Worst of all, the "crop dusters" squirt poisons on rich and poor, old and young. How could heli-choppers stir up more enmity?

This was the biggest spray war ever fought in urban USA. The stakes were high, the threat as grave as any in modern agriculture. It touched millions of lives, dashed political careers, and dominated the news for months.

Experts Shocked

To guard against flies, worms, pests, diseases, the state's Department of Food and Agriculture (CDFA) sets insect traps all over California, and inspects fresh produce incoming from all sources. So, in June 1980 when the medfly maggot raised its head in an apricot, it shocked the experts who didn't know this tropical insect could survive the cool Bay Area winter.

A medfly can deposit as many as 800 eggs – 10 eggs in one fruit – to hatch in a matter of hours. It lays eggs under the skin of no less than 256 various fruits and vegetables.

The fly is easy to kill with malathion, said Richard Jackson, a U. S. Department of Agriculture specialist. Spraying one acre with 2.4 ounces of malathion in 9.6 ounces of sticky bait lures the fly and kills on contact. Treatment in this war consisted of six or more doses, applied once a week.

Why did mere maggots create furor? Was it a "media event?" The war hit a political hot spot. Rich Santa Clara County is the heartland of the conservation-minded in the state with the strictest laws governing pesticides.

The Sierra Club, Friends of the Earth, and Citizens For a Better Environment, among others, denounced the state for spraying cities with malathion and other chemicals, and fumigating with ethylene dibromide – potent "EDB" which was later banned. They charged much of this was only useless, anyway, and areas were sprayed where no medflies were ever found.

Scientists took both sides. "We had more PhD's per acre than anywhere in the country," said Jerry Scribner, director of the project, "from the universities and over a dozen junior colleges. We had physicians who told the public spraying was safe, but we also had forty-four doctors sign an open letter saying one out of every 1,800 people would die or be seriously injured . . . ."

In wide areas the state ordered trees stripped of fruit; it released sterile flies by the billions, and ground-sprayed with chemicals. Workers treated 62,000 backyards six times in 1981. More than 4,000 fulltime workers fought the war on maggots.

Public resistance hit the ground spraying and the project had to keep records on 50,000 homes – which ones claimed medical problems, which refused entry of workers, which closed yards and locked gates or demanded special treatment.

The state and federal agriculture agencies quarrelled, though they don't talk about it now. The federal department (USDA) threatened to quarantine all California produce, which amounts to one-third of the Nation's production, unless the state began air spraying – as the one certain way to wipe out the medfly.

The blowup hit in December 1980. USDA wanted

action: fixed wing planes to spray the county from 500 feet above ground. Experts doubted the wisdom of such action, as medfly might only go dormant in winter, if it could survive. There was an outcry against the danger of low flying over cities.

"USDA," said one critic wryly, "knows about forest spraying, large tracts, but it has no experience with residential areas." Every city and county official vetoed the plan. The state expanded the medfly fight on the ground.

The war constantly fought in court, as it was involved with thirteen major laws and thirteen government agencies, plus the White House, the governors of Texas and Florida, and officials of Japan, Korea, Taiwan, and Mexico.

For self-protection Florida and Texas quarantined California produce, as did other government units. California even took the other states into the U. S. Supreme Court.

As head of the project, Scribner said he had to ". . . please a Democratic governor, a Republican president and state senate, and a Democratic state assembly all at once..." as well as officials of forty-four cities and eight counties, plus 2 million families.

More Worms Show Up

Always there were those maggots. State traps checked for flies in every square mile. Project leaders thought they were winning the war as they found no male flies, no larvae, for months. Suddenly, on June 24, 1981 the worms began showing up far and wide.

A stunned Scribner explained, "The textbook on eradication is fuzzy; each insect in a new environment is, by its nature, an unstudied phenomenon." CDFA changed tactics again and again. At Los Angeles the year before, they wiped out medflies with sterile males (to stop reproduction). But this year the method failed, possibly because some of the flies they obtained were not sterile.

Curiously, the infestation had little direct effect on the major growers. It was worst in and around the cities, mostly in backyard trees. But it still threatened a national quarantine of California farm produce. The big growers lost some $73 million to foreign and local quarantines.

Rather than spray, Governor Jerry Brown refused

Evergreen's Bell 212 on spray run; medfly war was fought at night over 1,300 urban square miles of California.

knowing advice and expanded the ground war. He had a tough choice: either spray his supporters along with everybody else, or let the medfly get out of control. He chose to risk the wrath of Agribusiness men, who had no use for him anyway.

Thus, the crisis was forced. And then, when he could no longer avoid it, Brown okayed helicopter spraying. Meanwhile, the deadly bugs had spread all over Santa Clara Valley.

Public meetings brought out irate citizens. They feared chemicals on food, helicopters over their neighborhoods to disturb the peace and privacy, crashes and fires and spills of toxic materials. A few extremists feared the rotor-beast more than the medfly. And the tiny insect went on spreading, even to Los Angeles and to Stockton, in the San Joaquin Valley.

Most of the opposition, Scribner found, centered on the nature of aerial spraying, which blankets everything and everybody. In urban places, about eighty percent of the spray would reach streets, roofs, parking lots, buildings, places where no flies could exist. Worse, rain would flush the stuff into drains and keep it in the environment.

Even Casper Weinberger got into the fracas. For unknown reason, the Secretary of Defense, an area native, forbade civilian helicopters to use the convenient Moffett Naval Air Station, in Mountain View. He relented, some days later, however.

Scheduling was another nightmare. Spraying in the morning brought anguished cries from dairy farmers concerned about cows; noontime angered the joggers and the luncheon crowd; midafternoon alarmed school officials. This forced the decision to spray between midnight and 6 a.m., hours when most people are indoors, animals at rest, cars put away. It's a time, too, of least traffic in the air, when the mild night winds have the least effect on spray drift.

Drinkers created a hazard. "Police protected our helicopter ground guidance crews," Scribner said. "More than 100 drunk drivers were arrested at roadblocks. Many of them hit the roadblocks."

Anonymous letters threatened the Project, and real bullets struck the helicopters at least six times. "One night I got a hole clean through my tail boom," said one pilot. "Fortunately, it didn't hit anything vital."

Liability insurance was another item of contention. USDA forced the spraying but would not take responsibility for damage to persons and property, and neither would the helicopter owners. The state finally accepted the risk.

For all the furor, the spray was extremely light: 2.4 ounces of malathion mixed in 9.6 ounces of syrup bait, applied at the rate of 15 to 17 pinhead-size droplets per acre — hardly like a deposit of dew. It baffled residents, as few could even detect where it fell.

The chief mechanic of Evergreen Helicopters, one of two spraying companies, said they worked a fine balance between fuel load and bait load. "Usually," explained Ed Bridges, "we gave 'em a load of bait and fuel for one hour of spraying plus the ferry time from our base and return. We could fly up to 2 hours, but with a smaller load."

But the sticky bait plugged their fine spray nozzles and the back pressure burned out pumps. Mechanics worked overtime to modify the apparatus. Meanwhile, Bridges said, spraying had to continue, so they burned up $20,000 worth of pumps before getting replacements.

Sprays Repeated Weekly

Weeks later, workmen found a ten-pound pump imbedded in the roof of a house. Pilots theorized that the discarded pump fell out of a helicopter.

The twelve helicopters worked seven nights a week, each night on a different corridor, but did not fly in rain, fog, or strong winds. They repeated sprays once a week until the CDFA certified the strip free of medflies — usually after six or seven treatments. The task began with 5 square miles, and grew to 1,300 square miles.

Evergreen's project manager-pilot, Darrell Ward, complained of the weather, ". . . always the hazard of sea fog around San Francisco Bay." He said this was one of his worst jobs in 21 years of such flying.

"Those runs were over 10 miles long; one was 23 miles. They were 6 miles wide. We flew echelon formation [five in a slant line behind the leader to his right or left]. At the end of each run all six would shift while turning back for the next run, with the lead and trail pilots trading place. That saved time and fuel."

With 12,000 flying hours, Ward is an expert agricultural pilot. While in high school he started in fixed wings, trained as a mechanic, and changed to helicopters in 1966. He was drafted for Viet Nam Army service but preferred to serve on the ground — "didn't want to fly the Army way."

Like most civilians he disavows Army slang such as "chopper," "helo," or "whirly." It is always "helicopter" or "copter," and he sees military flying as totally unlike this work over San Jose. The civilians have concerns such as insurance, which is unknown in armies.

Ward said civilian rules heavily constrained the medfly war. "We made elaborate preparations for those residents. The FAA puts strict controls on certification and maintenance. Army has none such."

The work was arduous, and the hours long. Ward took a rush call to Los Angeles, 300 miles south. "We left at 3 p.m., flew down there, sprayed from 10 p.m. to 1:30 in the morning, and came straight back to our group at Hayward Airport.

"I was only sitting in the pilot seat, but 21 hours on the job, 14 in the air — can't do that too often."

"We couldn't keep a neat schedule anyhow, midnight to 6 a.m.," he added. People had different squawks down there, so we varied a lot to please them. Early and late. We flew anywhere from 7 p.m. to 7 a.m., even in daylight. Had to get it done, once we started."

A soft-spoken, earnest man of 42, Ward said the work "was not too bad, though the gun threats made us mad as hell. Must have been 'one of every kind' down there. Maybe the medfly gave them a chance to let off steam."

Evergreen Helicopters, one of the largest in the business, has more than 100 helicopters of many types, plus a fleet of 25 or more fixed wing DC8's and other transports. For this task this company and San Joaquin Helicopters flew Bell models 205, 212, and 204. They chose these machines for their large payload and rush availability.

The Bell UH1 "Huey" series were made for the armed forces in thousands — the famous machine whose profile and slapping rotor noise made the daily television news in Viet Nam warfare.

Two-blade rotors on top and tail make the whop-whop noise. First produced in 1955, model 204 has 44-ft. rotor blades, 1,100 HP turbine engine. It carries ten persons or 3,000 pounds of cargo at 120 MPH.

A 1,400 HP turbine powers the 205. It has 48-ft. blades, carries 4,000 pounds, or seats fifteen, cruises at 130 MPH.

Basically, the 212 is like the 205, with the extra safety of two 900-hp engines.

Spurred by USDA, the CDFA in August added fixed wings to beat the fast-spreading medfly. Globe Aviation of Arizona flew four old DC4 airliners and three PV2 Navy patrol bombers of the 1940's. They worked the sparsely settled areas, mountains, and farm counties.

The DC4s delivered up to 1,500 gallons of bait mix on an 840-ft. swath; the PVs, 1,000 gallons on 420-ft. swath. To curb noise and safety hazards, they flew only in daylight, 1,500 feet above ground. Project officials warned residents that they might fly as far as three miles outside of spray corridors, for turns.

Copters generally cost three or four times as much as fixed wings. The two compare less easily than horses and cows. The two aircraft do different things, e.g., safety over cities, control of drift, speed, downwash to reach vegetation.

For instance, the fixed wing flies faster but loses time because it must operate from a distant airport; it has to move far out to turn into position for spray runs. The helicopter can do the whole task within the field boundaries.

Whatever the merits, nobody believes helicopters come cheap. The operators defend their charges however. Ward said simply, "Evergreen brought over $8 million worth of equipment to that job, and a big crew to run it on a *real* rush job."

Airplanes Not Selective

Of the heavy, 4-engine DC4 he said, "Airplanes have to go too fast, can't work small areas. They go over the cloud deck, not under. They can't skip places such as hospitals and reservoirs as we did. Like hitting a tack with a sledgehammer."

If they are, as some claim, a "breed apart," ag flyers have strong words for a public that persists in mistrusting them. With families of their own, they share the environment and insist they care for it like anyone else. They love to cite examples of disease curbed, starvation alleviated, lives improved by their work.

Pilots know the helicopter has enemies. But on soft July nights it served to put down a mean threat — as it always does in rescue and emergency. They note, too, when people are forced to choose, they favor chemicals over insects.

People dislike medflies as universally as hornets, snakes and rats. But few realize that with widely different habits and life cycles, the various bugs require widely different controls. Wiping out one pest may even help another, worse one.

Ag aviation endures strict scrutiny by law and regulation – by the Federal Aviation Administration overlord, and by federal, state and local agencies concerned with farming, pollution, business, labor, chemicals, and industrial safety. This means the nuisance of rules, inspections, restrictions, and reports, with fees to match.

The feds say one thing, the state another, the county something else. They never get together. Such is their common complaint.

These specialists prefer "farm flying" to the hackneyed, "crop dusting." Some use the more formal "aerial application." At best, it is crowded, competitive field of laboring early and late, on tasks dictated by bugs and plants and weather, and the wishes of growers.

The perennial hazards of flying they take in stride. The medfly war had its share of incidents, of malfunctioning copters and spray equipment. Pilots landed in open fields, parking lots, and once in a horse corral, but those ended without injury or damage.

In the one serious accident, Evergreen's Don Faircloth, flying solo, struck a house in Fremont. The midnight crash killed him and caused minor injuries to five people on the ground; it seriously damaged a house in the resulting fire. His spray tanks were empty, so there was no risk of spreading malathion.

Sea fog caught Faircloth, as Ward related the accident. "On our way back to base, suddenly we were in it. I was on the radio with him. I think he got vertigo – he snagged a power line. The rest of us managed to get down OK." In the fog, they scattered and landed at various airports and open fields.

This single tragedy in more than 7,000 hours of medfly warfare illustrated, as well as anything, the promise of aviation safety and its risks – one accident in a whole year despite the continual difficulties of flying at night in marginal weather.

Helicopters began farm work at their commercial start, 1946. Four decades later they are in demand for new growing methods, new foods and fibers, and new chemical aids for growers. They do countless chores. Although it is the costliest tool of farming today, the copter is efficient in controlling spray drift, in moving low and slow.

Scribner, a lawyer who once battled crusty bureaucrats on behalf of the poor, came to the job skeptical of helicopters. He is philosophical now.

What would he have done without helicopters? He would "make do" with planes, he says now, and deal with trouble they might cause.

"That Huey especially had a bad name — from Viet Nam — and here it was, flying over Our Town, in formation! But the fixed wing would've meant flying over 2 million people's heads at once. With this we could fight on small plots, localize the infestation."

Afterward, "the indexes that we watched went very well. People's anxiety fell off because we could assure them." And he recalled a Florida medfly infestation in 1956. Half the state, Miami included, was sprayed with malathion at 1.2 pounds per acre, *eight times what California used.* It had few adverse results.

Like most pest fighters, Scribner speaks of eradication as complete and final riddance. The fruit fly is ever ready to return and live on everybody's produce, which it will do, without constant vigilance. So far, science has devised no real extermination, no way to wipe this pest, nor any other, off the Earth.

Malathion did some damage. Game wardens found dead fish in a creek. An entomologist conceded that the poison kills friendly bugs — bees, beetles, lacewings, and parasitic wasps. Without the friendlies, secondary pests thrive and threaten growers.

CDFA technicians studied spray results. Although pilots turned off pumps over certain areas, the poison fell where it was not wanted anyway, though in small amounts. Spray drifted widely; in places they found up to 70 droplets per square foot.

The study asserted that science knows too little about long-term effects of malathion on environments, especially repeated exposure at low doses. The state senate blocked funds of $875,000 for a proposed survey.

No Sign of Spray Sickness

Although passed by the Assembly, the bill got nowhere. Newsmen blamed powerful agriculture interests. The action came after the State Health Department found no sign of acute illness from malathion following a study in the first months of aerial spraying.

"Opposition," said Scribner, "was unwise, and further tarnished agriculture's image with an urban public. One doesn't need to look further than the recent (national) EPA scandals to recognize that concerns about chemical use are widespread and legitimate, and should be treated as such."

The Assembly voted $3 million for property damage claims. The acid bait pitted auto paint, plastic solar panels, and skylights.

Although more than 16,000 claims (over $2 billion) were filed, Scribner said less than 100 alleged health damage. "I don't believe the state has paid any significant amount for health claims (aside from a few minor allergic reactions) primarily to the corn syrup, not malathion."

University of California computed growers' losses at $16 million, though few were directly invaded by the maggot. Nobody counted the loss of fruit that was stripped at small plots and private homes.

The project summary reported the medfly war lasted from June 1980 to September '82, 27 months, its final 56 weeks by air:

sprayed 2 million homes in 8 counties;
treated 6.4 million acres by helicopter, 3.6 million by plane;
stopped 5 million cars at roadblocks;
took 250,000 calls at phone banks.

Expenditures topped $9 million for staff pay, $2 million for ground spraying, $11 million for bug trapping. USDA spent $21 million more. The grand total, in direct costs, reached almost $100 million.

The state spent $18.4 million on air spraying. It paid Evergreen $7,722,379; San Joaquin $8,496,642; Globe Aviation (airplanes) $2.4 million. The helicopter owners charged $2,000 per flight hour, plus standby time — adding up to $144,000 for one night of spraying.

For the future, says Scribner, the state will "strengthen the external quarantines (ports of entry by land, sea, and air), to assure early detection of pests."

Since the medfly war the CDFA has used a different approach to such infestation. It sprays once, right away, then releases sterile flies to wipe out the pest.

We might expect a task of such magnitude to affect the populace, and it did. Scribner saw public attitudes shift. "They sprayed, the noise died and people asked, 'Is this all there is to it?' Nobody was hurt. Nobody died. After that we had little concerted opposition."

Residents today generally shrug off the episode as memorable bother. "We covered our cars," says Maxine Volstorff, who lives in posh Los Gatos. "We stripped our trees and covered the swimming pool. Our trees are back to normal."

Helicopters brought no furor. "We could hear them but the noise didn't keep us awake," Mrs. Volstorff said.

Mex-fly Also Found

Not long after medfly, the state discovered Mexican fruit flies in a 35-square mile section of southeast Los Angeles. Unlike the medfly, the mex-fly affects only citrus fruit, however. CDFA assigned four helicopters to spray malathion. The action drew intense press interest and public opposition. Late in 1984, another medfly stirred an alarm — this time a single one trapped in swank Beverly Hills.

Spraying, notably aerial spraying, has not quite won the day. Conflicts continue, because science has found little conclusive proof of its benefits.

Scribner warns against expecting the easy way out. He writes, "The experience is deeply etched in California's recent history . . . . Political leaders will tend to leave unpopular eradication programs up to administrators, so as to avoid falling into the trap that caught Governor Brown.

"The public, on the other hand, will be much more supportive of quarantines and exclusion programs and more willing to cooperate in pest detection and eradication — but not including aerial spraying. . ."

As he sees it, "the people will not welcome widespread use in their backyards and on their children and pets — based on government assurance that there will be no significant or long-term health consequences. The public knows we don't have data to support such assurances.

"The best that government can offer is evidence that it has not caused problems before, or has not caused problems in experimental animals. For a public dissatisfied with the way nuclear radiation hazards have been handled — and unhappy with the way other public health testing has been handled — these assurances won't be enough."

Will the medfly war repeat? Nobody believes so, not soon. This richest farming state controls pests with its quarantine stations, traps, and ag commissioners in each county. Perhaps more important, its powerful growers insist on constant vigilance.

In populated places live uncounted vermin, rats, gnats, ants, wasps, termites, cockroaches, etc., plus weeds and funguses and plant diseases. We shun them all, except when they prey on worse pests. Some affect health or safety or commerce, some pose trouble for appearance, but they readily adapt to environmental changes, or poisons. Even as the medfly did in San Jose winter.

5.

Food And Fiber

We share the land with 8,300 species of insects, 8,300 funguses, 160 bacteria, 250 viruses, 2,600 weeds.
— U. S. Department of Agriculture

In endless hordes they come, chewers, suckers, borers; with ingenious tactics and weapons they do their worst. The farm pests make off with one-third of what we grow, more than that in Third World nations.

It is continual and uneven human struggle against hordes that gobble up our sustenance. Worms, bugs, funguses, weeds, some are the more fearsome because we cannot see them with naked eye. They look innocent, but fight for every inch of territory, for every item we seek to grow to keep ourselves alive.

If this is war, it is the more deadly because the enemy has no feeling; whose countless mouths compete for what we grow to feed and clothe ourselves. We have picked them off by hand, set ingenious traps, fires, and campaigns; bent our knees and backs, used chemicals, sticks and stones, magic, germ warfare, elaborate machines, and prayer — and aircraft. After so much effort, eradication seems all at once imperative and hopeless.

Modern aircraft do a range of tasks, from tending and growing to defending crops, from protection of watersheds to collecting the harvest. And helicopters are used in various ways unique to field and forest. They sow the seed, spray defoliants and fertilizers; they aid in mapping, planning and survey, in cloud seeding to cause rain, watering in drought, and they lift baled and boxed crops.

More helicopters do agriculture work than any other task except air taxi and oil support. In the USA, fixed wings and copters together are critical to the growth of about one-fourth of the food supply.

Helicopter contribution to the world food supply is impossible to calculate, but it is substantial. We know ag products make up one of America's largest items of export, worth $40 billion a year.

Rotorcraft easily fly contour strips on hilly terrain. They keep constant speed up and down slopes for steady, even application on the crop. Having a loading site at the field means fast shuttle of loads and turnaround, and quick dispensing of material.

They can operate from a "nurse truck" carrying tools, parts and supplies, bins, and tanks of chemicals. The truck may haul the copter into the work field, and serve there as its operating pad.

Carrol Voss, a veteran of this work over four decades, wrote, "With full maneuverability in combination with the downward airflow through the rotor, and slow flight performance, the helicopter assures penetration and deposit on the crop.

"In his plexiglass bubble, the pilot has good visibility to do an efficient job of swath control, with reasonable safety. With short turns he need not climb over adjoining fields as he might with a fixed wing. That reduces mistakes and complaints. In addition, the farmer can ride over his field, beforehand, to show the pilot directly what he wants."

"The natural ag pilot enjoys his work. His philosophy turns on the idea of farming, of improving his work, of living it. He does not mind the pressures and the crazy hours that go with it, and the many problems."

Voss could so describe himself. He is a 9,000-hour pilot, owner of Ag Rotors, Gettysburg, Pennsylvania, an entomologist, agronomist, dirt farmer, as well as a respected authority on agriculture. He has a national reputation in agricultural aviation.

Voss knows airplanes well but declines to boast that copters perform better, although they cost more to buy and use. He prefers to stress "the good we can do rather than the deficiencies of other methods – ground and air, they all have their place."

Helicopters spray fertilizers and insecticides on large forest areas; this is Evergreen Helicopters Bell 205.

Ag Rotors Bell 47 spraying, inches about cotton crop.

Must Overcome Risks

It is the measure of Voss's definition that the natural ag pilot must overcome the twin risks of flying and chemicals. To serve in safety — to make a real career of it — he must think with an ungainly spraying or spreading contraption attached to his aircraft, in addition to the spinning rotors on top and behind him.

And he has to understand this involved machinery and the nature of the chemicals. Which means, all at once, precision with flying speed, droplet size, character of the chemical, volume, height, limits of the field, wind drift, temperature, density, humidity, flying obstacles — a long list of variables.

"Good ag pilots," Voss asserts, "must first be interested in agriculture. They cannot fake that. How could they communicate with farmers?

"They have to work hard like farmers, willing to put up with tough conditions, get their hands dirty, know how to fix anything, fly in the tightest corners — in hazards around trees and buildings and wires. Do all that without wasting time."

There are farmers and aviators without understanding. In its complexity, agriculture needs bacteria, insects, viruses, funguses and weeds, to serve one purpose or another. But they are pests when they invade our habitat, even our bodies.

These living things are friends when the vegetable grows wild, when bees pollinate fruit, when bugs we tolerate prey on those we oppose, and the fungus mushroom pleases the gourmet palate. Not all pests are enemies, nor enemies for all persons or all the time, nor enemies in all circumstances. But even the friendlies can overwhelm the growth we want.

We share the Earth with more than 8,300 species of insects, 8,500 funguses, 160 bacteria, and 250 viruses, plus 2,600 species of weeds. So says the U. S. Department of Agriculture. Billions of billions of living things, versus a few billions of humans. Enough to make us ask, who owns this old Earth?

To grow anything in quantity we have to keep the land fertile, protect it from erosion, storms, drought, and floods, as well as pests. There, literally, we earn our daily bread. That's where aircraft do their part. They deliver what farmers ask, where they want it.

And when they want it. Voss's schedule opens in January with clover and alfalfa seeding, weed control sprays for spinach, potatoes, onions, and clover. In March they do more, and add grain sowing.

They dust grapes, tomatoes, clover, and onions in April; then beets, peaches, cherries, potatoes, sunflowers, vetch, melons, and celery in May and June. Then come beans and walnuts in July and August, plus hormone sprays to stop premature dropping of pears.

Seasonal work in September means spraying cotton defoliant, seeding of grass and clover in October, weed and brush control in November and December, plus dusting of celery and peaches. Such is the work around rural Pennsylvania. For off days, the Ag Rotors pilots work at flying tourists, patrolling power lines, herding livestock, and flying photographers.

For Voss, hard work and many risks to person and fortune brought satisfaction in achievement. In his shop he grins over a 1946 four-wheel Bell 47B, made in Niagara Falls, New York, the third off the assembly line of this first commercial helicopter model. Tender hands restored it to original shape.

"In 1958 we bought our first 'ship,' one just like this. It had been modified with a bigger engine and a cut-down cabin, like an old roadster car, for ag work." We still use this one a little, for certain jobs." He restored another 47B, which has gone to a museum.

Today he has a fleet of eleven Bell 47G's, two Soloy Hillers, two Hiller 12E's, two 206B Jet Rangers, and a Robinson R22.

Satisfaction is reflected in Voss's friendly manner, bright blue eyes framed in horn-rim glasses. His close-cropped hair edges a mostly bald pate, with trim chin whiskers around a ruddy face. The combination gives him an owlish look, and younger than his 65 years. He appears less a farmer than a professor at a small college.

Voss's unimposing base of operations, ten acres of rolling farm land, borders the Gettysburg Civil War battlefields on three sides. His small but pleasant office has a comfortable clutter. He aims a deadly swatter at the pesky Pennsy flies — this entomologist won't overuse bug killers.

In the August heat tourists clad in shorts throng the town and the Civil War scenes that stirred Abraham Lincoln's eloquence. You hear talk of "rich country" and "good corn weather." Which it is, and it fails to daunt the pilots for whom it creates work.

Tourists dribble into Ag Rotors for sightseeing rides. Over the battlefields they fly at decent heights and routes in respect for the peace of the cemeteries. With his son Tim, Voss manages the business of thirty employes.

Carrol Voss is, foremost, a man of the soil. He sees himself as an entomologist with graduate degree in aviation. The Nebraska Future Farmers of America boy grew up raising pure-bred hogs, corn, and other crops, and got into higher education despite hard times.

"A $100 Sears scholarship did it, and how far that money went! I had a year at Lincoln Ag College; from there I went on to BS degree at the University of Nebraska, majoring in agronomy and entomology."

The Second World War got him into flying, in Navy patrol bombers. Conviction about flying turned him to helicopters after the war, while studying at University of Wisconsin for a PhD in entomology.

Although post-war times were lean, helicopter pilots were scarce. "I worked for a New England operator, making $40 a week, moving from job to job. What kind of a future was that?"

If nothing good ever comes easy, this grownup farm kid learned perseverance. He held many ag jobs, worked as a consultant, sprayed bananas in Ecuador, rubber trees in Liberia, sugar cane in Trinidad, other crops in India and Upper Volta. With a troupe of scientists he toured Russian farms.

Helicopter pilots do get around. It was in Voss's mind to learn all he could about pests and controls in the various climates and different ecologies.

Cotton Treated the Most

Of all U. S. farm products, cotton gets the most attention by aviation, about half of the acreage treated. Some ninety per cent of American rice grows with the aid of helicopters and fixed wings, on 3 million acres — most of it in Arkansas and California. It yields over $1 billion a year. Aircraft cover as much rice acreage in an hour as ground machines do in a day.

Rice farmers plant pre-soaked seed that is ready to sprout; copters or airplanes fly it at 15-30 feet above the paddies. They then drain the field, dry until it cracks, then spread fertilizer at 300 pounds to the acre. To enhance growth they flood the field once more, then drain again. As the rice grows up with weeds and grass, the aircraft come with weed killers, then the paddy is flooded for the season. They add more herbicide and fertilizer as needed.

Far unlike custom in Asia, aircraft replace coolie labor here.

The Swedish government fertilizes forests to enhance growth of wood. Bell 205 helicopters lay down the urea nutrient from a 600-gallon bucket carried below on a sling. The bucket carries up to 4,000 pounds of granules, 30-100 feet above the tree tops. At 85 MPH the bucket empties in 3 minutes.

Sterner Aero Sweden uses red-white-blue copters of an American firm, Heli-Jet. The buckets are fitted with quick-disconnect hoses and cables. Inside, a small motorized spreader shoots out the urea in a 210-foot swath. The bucket contains a 7-gallon tank of fuel for the helicopter too, thus maintaining a constant, light load of fuel.

In action, the copter hovers low while the ground crew detaches the empty bucket and hooks on a full one. Switching takes 15 seconds, then the copter speeds away. In minutes it returns for another load.

Rod Kvamme, president of Heli-Jet, said one helicopter "can spread 330 tons in a day. That covers 1,490 acres with urea. We do herbicides too, on a different scale. Of course, it's a big crew. When we move we look like a gypsy caravan. We have two pilots and a mechanic for each helicopter, plus four on the ground." Heli-Jet works its summers in Sweden, its winters in the USA.

The Swedes also spread lime to neutralize acid rain, thousands of tons of it, to protect fish life. Wind carries acid rain to Sweden from industrial Europe. This persisting trouble of modern times is a growing menace for much of the northern hemisphere.

Airplanes equipped for agriculture are not cheap — from $80,000 up, as high as $200,000. Helicopters cost about twice as much to buy and use. This hazardous work incurs high insurance rates; in addition, there are high insurance rates for the risk of chemicals misplaced or leaking in unwanted places.

Operators claim they apply two-thirds of the chemicals

used in agriculture but consume only ten per cent of the fuel of ground rigs. Agronomists assert that, without aircraft, food prices would rise at least fifty per cent. The National Agricultural Aviation Association estimates that aircraft account for $15 billion in food and fiber.

"Not every aviator can be a good ag pilot," says Voss, who grew along with a profession that opened in 1946, Year One of civil helicopters. In time he opened his Ag Rotors school for pilots at the south edge of Gettysburg. It is one of the few devoted primarily to this specialized branch of flying.

Here they teach the life cycles of plants and animals, bugs and bacteria, weeds, funguses, and parasites, the minutiae of apparatus for spray and dust, and flying techniques just for this purpose. The course, costing $7,500, includes three weeks of ground study plus 55 hours of flying.

For example, they learn how to dust and spray in calm or light winds, so chemicals won't drift off-target. Helicopters perform better in wind; it helps more than it hinders. Since lowest winds normally occur between sunrise and mid-day, or again at sunset, pilots must arise before chickens do, and they work late, like the farmers.

Of their weather bugaboo one pilot snorted, "It's like getting married. Always something against it, just now."

Of weather quirks, Voss said easily, "We have to live with them, don't complain. Farmers don't gripe that much. We might have to get up before dawn and wait, day after day, ready to go at first light. Patience works."

Third World Lacks Funds

If ag aviation has made its mark in the USA, poverty allows too little money for it in the Third World. The blood-sucking tsetse fly, for example, has ravaged primitive Africa for generations — a constant threat of sleeping sickness for more than 100 million people.

Without cure or preventive measure, the disease causes humans and animals to sleep unto starvation. The only known remedy is control by eradicating the tsetse.

This housefly-horsefly lookalike carries a parasite that causes the sickness. In proportion the tsetse takes heaviest toll among livestock. In Nigeria, oil wealth has enabled the country to carry out a helicopter spraying program.

Nigeria has cleared about 150,000 square miles of the tsetse and has seen a revival of the cattle industry.

The NAAA estimates that fifty-five nations treat 600 million acres by air. The USSR leads in this endeavor, with the United States a close second; the two do about seventy per cent of it, followed by Mexico, Argentina, New Zealand, and Canada.

Socialist USSR does not compare directly with Western nations, as its ag aviation operates on a different system. The Soviets fly more than 12,000 ag aircraft, servicing the great land mass on two continents. A scientist, V. A. Nazarov, wrote of it:

"By 1958 there was marked increase in the volume of airborne operations in combating pests . . . . An important advantage of the helicopter over the airplane was the better quality of pesticide distribution . . . . Helicopters have proved irreplaceable in combating vineyard diseases, various orchard pests, mites and the blood-sucking fly in the taiga forests."

"It is possible," he said of costs, "to obtain more annual production amounting to 3-5 billion rubles. Each ruble expended in plant protection results in a 3-10 ruble payoff in regard to cereal crops, 10-15 times on cotton, and 15-20 times on orchards."

The big USSR bureaucracy lays it on an army-type organization to serve the collective farms, state farms and other groups, with ". . . agreements, preparation of staff and aircraft, delivery of fuel and lubes to the site . . . . A schedule is to be drawn up for the command-supervisory staff to inspect . . ."

And then they "leave for the region of aerochemical operations 10-15 days prior to the departure of the aircraft." Nazarov's writing smells of rubber-stamp government serving itself — the aircraft just might arrive, when they get ready. You pray your crop is at its peak.

Still, Soviet operations are said to be effective. Helicopters of three or four types make up half of the fleet. They have good performance and payload, notable the twin-engine Kamov KA26, a reliable old model with a coaxial rotor system.

Locusts Plagued Russia

Soviet ag flying dates to 1919, when they fought off a plague of locusts. This notorious pest of history comes in cycles, often swarming in a cloud over miles of land. The locusts devour all vegetation and fibers in their path, including fences and posts, even tool handles, and leave famine in their wake.

The locust has not vanished. It plagues India, Africa, and the cycle shows renewed signs in the American West. The known pesticides kill both the insects and their predators. Then, with a small start, locusts appear in explosive hordes until the predators revive.

In the U.S., the gypsy moth classifies at least as our most visible pest. This not unpretty butterfly, whose American history dates back to 1869, has withstood all campaigns of eradication. It bounces back and spreads, ever farther — starting from Massachusetts, it has crossed to the Pacific Coast.

The moth caterpillar feeds on oak, elm, maple, leafy trees we like around homes and parks, in "good neighborhoods." When hungry enough, it eats anything green. That's the worst of it.

Not a destroyer like the locust or fruit fly, these larvae feed on leaves. They will strip trees clean and kill some of them. They too can appear in hordes, caterpillars crawling all over the trees, falling off like rain — an explosion of so many caterpillars as to make streets and sidewalks squishy slippery. Their own kind of hell.

Nobody knows how to do in this pest permanently, although spraying campaigns with a controversial chemical, carbaryl, or a bacterial agent known as Bt Dipel 4L, usually bring it under control. Spraying around residential areas, in small plots, means helicopters, of course.

Carbaryl kills bees and certain other friendlies. Dr. Voss favors Bt, sprayed twice on a given area, in small plots, but that pheromone is expensive and works slowly.

Voss won't be baited by environmental questions. He projects a middle-road approach leaning toward chemicals. "We need to know more about them, about our aircraft, systems to make them better, more about droplet size and combinations, and drift, for better results.

"We don't have a risk-free society. We must accept the blame for mistakes in ag flying. I know that. But it does not

Farm copters assist harvest, e.g., lifting Christmas trees.

mean we eliminate chemicals. We cannot do without them."

For a century the moth has frustrated every move against it, and has kept laying eggs in any old place — clothing, the undercarrige of trains, autos and trucks, lumber — to migrate to new habitat. Nobody has easy answers to this pest. There are none.

The perennial argument: is this pest bad enough to take action? Spray? Chemical or other means? Ground or aerial? Helicopters?

In two recent instances, the screw worm and the fire ant once threatened the southern states. Chemicals brought them under control, just as, in history, science learned how to control the carriers of malaria, the plague, yellow fever, even the tsetse.

We do have alternatives. The best answers of the experts add up to constant care to control the bugs that insist on sharing this Earth with us. We think we own it, but who can even count *them?*

The worst of it, some of these little creatures prey on other varmints that are our allies against the bugs that bug us.

Their history foretells of more pest warfare to come, probably on increased scale just to keep up with them. How to get rid of those bugs? Or funguses or "weeds"? Or fertilize to produce more?

For a few billion human mouths how much chemical control must we accept? When forced into action, how best deliver?

Carrol Voss spoke for agricultural aviation when he said, "With all the technology devoted to progress by chemists, entomologists, engineers, and others in developing the aircraft and the treatment formulations, the distribution equipment and the special techniques, in the end, the result comes into the hands of one man, the pilot."

The grubby ag pilot working in the world's backyards, living and dealing with toxic stuff aloft, gets no public thanks or recognition, but he is the front line fighter in an unremitting war.

6.

Yours For Public Service

The helicopter brings the police presence in seconds.
— Chief Lee B. Brown

"I was just cruisin' when the city police put me onto a 'cycle' speeding through town. I picked him up quick — the one matching my airspeed down on the freeway. It was reading 90."

John Saucier, a part-time deputy sheriff pilot in a helicopter answered the call of the Slidell, Louisiana, cops. He tells his mousetrap story in bayou drawl.

"I tracked him maybe 2 miles when I saw he'd go through this freeway underpass up ahead (like a tunnel), so I just slipped over it and down to a low hover, about 50 feet over the road.

"He came roaring out of that underpass, spotted me, and ground to a halt. Then he hopped off his 'cycle and waited for the patrol car. He was standin' ready with his driver's license in his hand.

"He gave up just like that. He said to the patrol officer, 'Where in the hell did that copter come from!' "

Saucier's story belongs to the lighter side of public service flying. Most often, law enforcement aloft consists of long periods of cruising, wondering what will break the monotony, and when — always dealing with what-ifs.

Take a ride with Police Lieutenant Paul Newell, Columbus, Ohio. On a midnight patrol in the Hughes 300C, he is over an industrial district when the dispatcher sends him tracking a carload of burglary suspects. He noses the helicopter down to 300 feet and hauls back air speed, while the observer switches on the NightSun.

Instantly, the 3 million candle-power floodlight shows a stark scene of roofs and streets and building hulks. Instantly the two officers pick up the car speeding without headlights through a factory gate. The observer mutters, "Now, where can they go!" In this glaring-white-midnight, what can we miss below?

The air officers keep a steady patter of directions as they track the getaway car. Our earphones tell us patrol cars are hustling from three directions. We eye their flashing red-blue beacons; one covers 7 miles in 6 minutes. In city traffic.

From aloft it is theatre madness, a fast tableau in the vertical dimension. Cars rush down empty streets, three converging on one; they careen turns and crash red signal lights. We imagine loud noises and squealing tires but the clattering copter fills our ears.

The burglar car runs headlong, its pathway aided, ironically, by our floodlight. No matter, this is a dragnet with lights.

One patrol approaches head-on, another from behind; the burglar probably sees both. He seizes a way out, darts down an alley, right into the third patrol car. Neat trap.

It seems unreal, a TV drama. In reality, four cars clustered, a Christmas-tree of flashing lights, figures darting about, officers with guns drawn, two men with hands high being shoved face-down over their car trunk, police handcuffs. "Ten Four," says the radio. NightSun off.

In 17 minutes it is all over. Newell wheels up in climb, back to the beat. Now what, humdrum cruising in race track course, or next midnight adventure, which? More likely, says our pilot cop, just another two hours in the dark.

They Seldom Land

Not much to it, Newell remarked later. They can land to work on the pinch, but seldom do. The copter does best when it's aloft.

So why do burglars persist when they know, as they sure must, that from the police copter is no escape? Columbus has had this patrol for years. Is it a gimmick, a high-tech

In Hughes 500C, Houston police deliver instant response to calls.

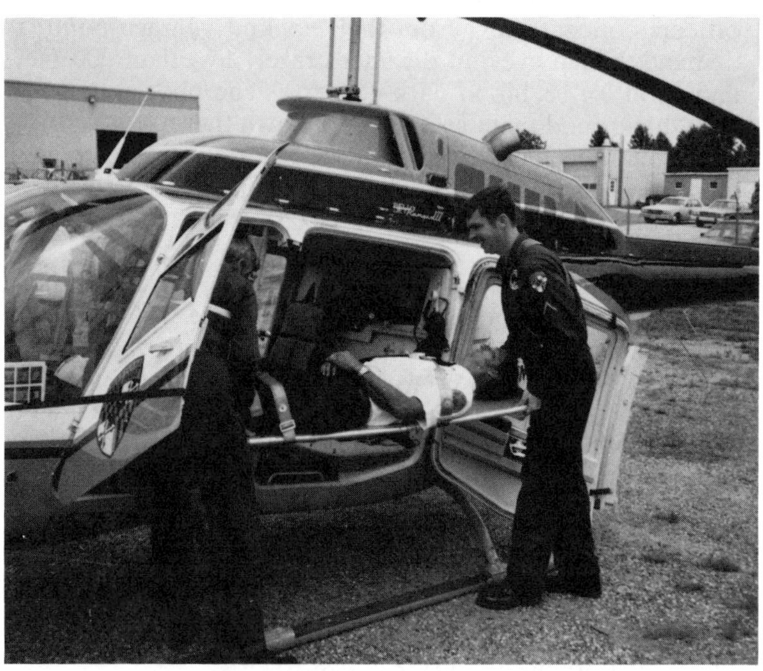

Maryland State Police cover entire state by copter.

plaything of the spenders? What of the cost, the risk of crash, of being shot down?

What if the burglars used a rifle on that NightSun target? It did happen in Viet Nam and in civil flying; instant blackness would blind police eyes aloft when needed most, during the getaway. Ohio has notorious thunderstorms; what if weather interfered — rain, haze, nasty lightning, fog, icing weather, high winds?

When seconds count, what if no helicopter were up there? What if all nearby patrol cars were "out of service," pre-occupied with other duty in public safety? In that case the copter could stay on chase and summon the sheriff or other police — or track down the quarry and land for an arrest. But, in time? Police answers center on timing as prime element. Along with the combination of expert police work and piloting. What if . . . .

Newell is a veteran of war service and the Columbus air unit; he's instructor, officer-in-charge, safety officer, check pilot. Good police work, he says — no game for amateurs. The Thin Blue Line has to do the best in courage, judgment, and skill. In air patrol that means close teamwork, two officers concentrating from their cockpit vantage point; it demands sharp eyes, at distances rarely less than 300 feet, and good for seeing in dark corners. The pilot must work with the ground team as if he were down there.

Columbus has a fine safety record; no serious accident in nine years. Just last month, we remember, Patrolman Bob Cormier died in a Dallas, Texas, crash along with an observer.

"Bank robberies, car chases, accidents, we've had them all." A big man, though not your burly police type, Newell is lean and trim. "We fly about 2,500 hours a year, and we miss very few days in winter weather. Night patrols are most effective, 8 p.m. to 4 a.m. Our average response time is a bit under 2 minutes. Our best record, in one day our helicopter patrol took part in thirty-two arrests."

On an August morning the patrol unit made a "real bonanza" arrest when a civilian landed suddenly, without permission at the downtown police pad. He brought a man who was writing bad checks. Pilot Rob Case became suspicious when the man, who turned out to be only 17, telephoned to charter a ride across town.

The youth did not object when Case quoted the charter rate at $300 an hour, and said he had just come from

Cincinnati. Case called the airport and learned the juvenile had passed a bogus check. After he picked up the boy, he used an excuse to land at the police pad.

Federal funds got Columbus started in 1971, with two small Hughes 300's and seven officers. Four years later the unit had its own new heliport in the heart of the city, with complete facilities for all needs. After another three years this branch of the Thin Blue Line had to fight for its life in a budget crunch. Today it consists of five helicopters, ten policeman pilots, and seven uniformed observers.

Law officers need an awareness of the criminal mind. Why don't burglars think of that helicopter? Or do they think? A police axiom says burglary runs small risk and looks simple, so it appeals to lazy guys out for the easy take. Most of them net small loot, those who "break and enter" warehouses, stores, taverns, private homes, garages. Yours and mine. They get caught sooner or later. Police scorn burglars as dummies.

Maybe so, but the lawmen rarely talk about the experts, the gentlemen pro's who knock off bank vaults and jewel cases. Too often they get away with real big money.

Combine the demand for expert piloting, fast thinking, alertness, a linkup for close teamwork with ground patrol, plus an array of electronic aids. The public service helicopter carries the usual tools of aircraft and of law enforcement. It has its own guarded radio channels, large floodlights, loud speakers, and first aid and rescue equipment. So there is much more to it than just putting the policeman aloft.

That vertical dimension changes the rules and the odds. By one measure, two men in a helicopter can do the surveillance work of thirty-five patrol cars. And respond to any call three times as fast. It is not that simple, of course – some things helicopters cannot do, viz., fly in fog or go inside a tunnel or a patch of woods, or land on every wire-cluttered street. Their time aloft has limits. Fog or bad weather can seriously hamper the vision of the patrol

Across the United States, some 500 jurisdictions – mostly cities, counties, and special districts, a number of states – operate about 1,500 helicopters. Aside from chronic financial problems, mixed or weak public support, modest or battered equipment, and a certain defensive eagerness to prove themselves, few of the helicopter units have much in common. Because no two cities, counties, or states are identical, they operate under different rules and practices.

Police throw numbers around, but they cannot prove the value of helicopter patrol in dollars and cents. Units have blossomed, served well, then got the axe for failing to convince the people who pay for it. In Chicago and San Francisco, accident or political factions killed them.

'Essential of Law Enforcement'

In Houston, Texas, Chief Lee B. Brown sees copter patrol as "the essential of law enforcement." There, the police copter fleet covers a fast-growing metropolis of 600 square miles. Two are on patrol at all hours (except 4:30-6 a.m.); four stay aloft in high-crime periods of afternoon and evening.

On a basic budget of $3.3 million, the unit flies 11,000 hours a year. After fourteen years with this system, Houston's Finest have little trouble justifying their budget. It is "not appropriate," Brown said, to measure the value of helicopter patrol on arrest rates. "We have a multi-million computer system and we don't justify it that way either."

"We provide services (a few hours a week) to other city agencies, but our main emphasis is law enforcement," Chief Brown said. "Its best attribute is response time. The helicopter always brings police presence within seconds."

"They usually beat the patrol cars — in as little as 2 seconds. Their average response time is 2 minutes."

Under Captain Ellis Milam, the Houston unit flies six Hughes 300's, five Hughes 500C's, and hopes eventually to convert to all turbine helicopters (like the 500). The staff includes sixteen pilots, seventeen observers and mechanic and office personnel located at Hobby Airport. They work on regular pay scales — no extra pay for flying. Like others on the force, they get no overtime pay either.

Captain Milam cited one of the most novel cases, when a man hijacked a police car, with "beanie" lights and sirens blaring, and drove at high speed through downtown streets. Meanwhile, the copter patrol tracked every move and kept three patrols advised.

When the hijacker rammed a building and jumped out, the overhead patrol knew his hiding place so ground patrols could pounce on him.

"One of our worst problems," Captain Milam said, "is the low air speed of the 300. At 85 MPH it cannot buck head winds well enough, and we get strong winds off the Gulf (of Mexico), or, in winter, those 'Texas Northers'."

"Six hundred square miles is a lot of ground to cover for only two copters against head winds. We get over 22,000 calls a year."

"Sixty per cent of those calls come in night hours. Then we have to use the turbines (Hughes 500's); it's better not to fly the piston engines at night. Of course, we have to be in the air to do our best."

"We cannot respond well enough from the ground – it's already too late for takeoff after a call comes in."

Houston pilots can land and make an arrest, but rarely do. That, said Milam, is up to the pilot. In the main, the pilot flies where the observer directs him.

Said an observer, Patrolman L. J. Faultry, "The most satisfying thing to us, when we get on a chase, is to know they can't get away. Up there I often feel like Pac Man!"

Surveillance, one of the prime capabilities of the helicopter, can be abused, as police critics charge: "Snooping, violating the privacy of innocent people in the name of law enforcement." Are helicopters used for that purpose?

"Never, here," responded Chief Brown. "We have plenty to do without that. Preventing 'snooping' is a matter of supervision. We are busy going after criminals."

Brown was brought to Houston from Atlanta, where, as public safety commissioner, he supervised the solution of a notorious series of murders of young black boys. He holds a PhD in Criminology, and established the Division of Administration of Justice at Howard University, Washington, D.C.

Federal Air Arm

In Washington, D.C. the federal Park police work hard at rescue (as in the famous 1983 Air Florida crash)[1] as well as in unseen actions, and are well regarded. The District of Columbia police, who do only criminal work, often must fight for operating funds. Because of strict curbs on aviation, the DC police must fly out of National Airport – across the river in Virginia.

The park police work as a kind of federal air arm – often an aerial command post – quietly aiding the Secret Service and other federal units on demand. Based at the Anacostia River in DC, they are virtually the only helicopters

[1] See chapter 2.

permitted to fly in the Prohibited Area over the national capital, other than the President's own Marine Corps copters.

Normally, park police copters stay over the rivers enroute to and from their home base. "Our patrol job," Police Pilot Earl (Butch) Cronin says, "is broad. We look for anything out of the ordinary." That means rescue from accidents about forty per cent of the time. They have moved the injured to hospitals from as far as 50 miles.

The New York Port Authority had the first civil helicopters in 1948. They do surveillance, transport, and rescue. New York City (NYPD) initiated police copters about the same time. They proved so useful that by 1954 the police retired the last of their fixed wings.

NYPD copters patrol beaches and boat traffic, plus 500 miles of bustling waterfront. They monitor traffic on streets, highways, and bridges, and check all airports within the city limits. They fly from the old Floyd Bennett Field in Brooklyn.

Given the rocky state of community relations, the growth of violent crime, and tightening city budgets, it is remarkable so many police fly in the USA. Still, given criminal behavior and the effectiveness of the air patrol, it's a wonder so few of them fly.

Public Service copters have totted up a controversial record. One police commissioner called them "a fad that grew out of federal funds." Still, though such funds are gone, crime is not. As population grows, so grow demands on law enforcement. And city budget makers contend as best they can in the face of intense opposition and the outcry for tax cuts.

Somehow, the public service units hang in there, with reduced flying hours or personnel when they have to. In the past decade the Airborne Law Enforcement Association noted a slow but steady growth: "at least sixty-five new units added."

Baltimore, an industrial city with one of the busiest seaports on the Atlantic, concentrates its patrol on the high-crime core area. The helicopter can reach any point within the city's 89 square miles in 90 seconds.

Police in a western city keep a fixed wing Cessna hidden in their hangar. At irregular times it patrols at 1,000 to 1,500 feet altitude because, its commander said, "from interrogating prisoners we learned they hide when they see our helicopter out there, but the light plane looks innocent. Sure enough, we've turned up many a suspicious activity with that plane, then we send out the helicopter. Keeps our arrest record up."

"They don't know we have it. Don't you tell 'em!"

The Volusia County, Florida, sheriff has a fleet of helicopters and one airplane; he got a new Jet Ranger through confiscation of a drug smuggler's property. Volusia pilots fly surveillance, find lost persons in the coastal swamps, work traffic, accidents, rescue, and marijuana busts.

"We do many tasks for the county government," said Sgt. Terry Sanders, deputy-pilot, "though we don't fly politicians. The County Manager won't let them interfere with police work. We use our plane to move prisoners in extradition, and sometimes we can drop him off at Washington D. C. enroute. That way he saves the air fare."

"It costs less to fly prisoners ourselves than use the airlines, you know. Usually it's out and back same day. Airlines require two escorting officers (who get $50 a day per diem) and won't take violent prisoners at all, nor let us have guns. They don't really want us. Such an airline trip can cost us $2,000."

Sanders described Volusia, on Florida's northeast coast, as "60 miles long, 35 miles wide, and 42 *feet* high." Its largest city is Daytona Beach. County population of 268,000 swells to 1 million and more in tourist seasons. Of course, such crowding of sun-seekers tends to multiply traffic, crime, and accidents.

Response Time Vital

The Phoenix, Arizona, police concentrate on response time. "Our arrests," said Sgt. Tony Kruczynski, "go down whenever the response time goes up."

The clincher for the Phoenix unit came early in its history when the flooding Salt River inundated this normally parched desert city of 1 million. Arizona thunderstorms are prodigious. The air unit went on 12-hour shifts for ten days before the river receded. All other means of travel were blocked by flooding or washed-out bridges.

Helicopters provide the vertical dimension for many a sheriff, state police, park ranger, game warden, and patrol force throughout the Nation.

The Maryland State Police cover mountain, city, and Chesapeake lowland, border to border. They base helicopters in four sections, so they can reach any place in the state within 20 minutes. This Thin Blue Line has prospered since it began in 1970.

At the start, Captain Gary Moore said, "our biggest problem was public awareness of what we could do." To make a reputation, the unit sought opportunities in search and rescue and accident, and created a close working arrangement with trauma centers in Baltimore. Now, "some forty per cent of our flight time goes to rescue and accident."

The City once called on the unit to sling-lift[2] an airplane from the bleachers of Baltimore Memorial Stadium. A drunken pilot had crashed there in an attempt to land on the short infield.

In another incident, a burly pilot, with an observer who was just as heavy, landed to help a man injured when his car struck a tree in rural Anne Arundel county. The victim weighed 320 pounds — "so fat that we had trouble getting a protective collar around his neck." His injuries were serious. They saw it would be difficult to put him in the helicopter.

The crew faced a worse problem. In this muggy 102-degree air, the victim's weight plus weight of the pilot and observer, would the Jet Ranger even leave the ground? There was added risk of spinal injury, so they decided to get the victim to the hospital in a ground ambulance.

Again, one dark night in 1973, in a Jet Ranger, MSP Trooper Don Sewell chased an Army Huey helicopter, without lights, at Baltimore Friendship Airport. When the Huey landed in a rural area (the pilot bought a six-pack of beer), it was recognized as stolen.

As the outlaw began to take off, a police car rushed up. The Huey pilot knocked one of the beacons off the police car with his skids, and sped off toward Washington, 40 miles away. Sewell was in hot pursuit.

The Huey headed into the Prohibited Area at 500 feet, hovered over the Washington Monument, then hopped down to 5 feet above the ground. He turned on all his lights, buzzed the MSP copter, and went into the White House grounds. As the two copters hovered nose-to-nose, Secret Service guards shot into the Huey cockpit.

"We had this new Jet Ranger — only thirty hours on it," said Sewell. "I was afraid we would smash rotor blades, nose

[2] See chapter 9.

to nose like that. We were afraid for our lives. Did that wild man have armament, more firepower than our 38-cal snub-nose guns?"

Finally, the Huey pilot surrendered. He turned out to be a disgruntled Army sergeant who had a civilian pilot license. An Army court martial gave him a jail term and a heavy fine, and he lost his pilot license. Trooper Sewell and his observer received Congressional commendations and a White House ceremony by President Nixon.

California Concentration

More helicopters fly public service in southern California than anywhere else. Small cities and large ones have their own, altogether about 125 helicopters.

Naturally, the farthest flying of these air units is in San Bernardino County, 20,064 square miles of desert land lying northeast of Los Angeles. In area, this largest county in the Nation exceeds Connecticut, Massachusetts, Delaware, and Rhode Island together.

Sheriff Frank Bland's Aviation Division covers this territory with two Hueys, four Hughes 500C's, one Hughes 500F, and five fixed wings. Moreover, the Division furnishes local patrol, at nominal charge, for nine cities and unincororated areas, in a so-called regional patrol. It is based at Rialto Airport.

The unit's staff of twenty-two (nine pilots including the commander) fly a total of 7,000 helicopters hours a year, and are involved in 2,500 arrests and hundreds of rescues.

The Division primarily is the creation of Captain Terry Jagerson, 46, a zealous promoter of the public service cause. For all its size, the county has only 900,000 residents, most of those in the 1,000 square miles of its southwest corner. Said Jagerson, "People throughout the county have a right to police protection. We can give them that — at lowest cost — with helicopter patrol."

"This is about the biggest recreation county in the country," he added. "It draws the RV's, motorcycle clubs, ultralights, hang gliders, all kinds of outdoor pursuits."

The county is crossed by freeways to Arizona and Las Vegas, but much of it is untracked desert wilderness. It features mountains as high as San Gorgonio, 11,850 feet, and lowlands under sea level in a corner of Death Valley;

old mine shafts, dry lakes, blue mountain lakes, and winter sports area. The climate has extremes of cold and heat year-around, and storms of hurricane force. There are accidents, flash floods, forest fires, road washouts, and some noted manhunts.

The desert beckons the thrill-seekers, the criminal, the careless — shooters, hunters, desert fanciers. The helicopters, said Jagerson, have discovered hiding places unknown to the outside world.

Jagerson favors the Hughes 500 helicopter because of its reliable turbine engines, its four-five seat capacity, 140 MPH speed, and small (27-ft.) rotor diameter, usable in tight landing spaces. He took on the Hughes 530F because it can hover as high as 13,000 feet in ground effect[3] — useful in montain rescue. His fixed wings transport executives, detectives, prisoners, and cargo, and make searches in the wide open wilderness.

Jagerson claims that "regional patrol" costs small communities less than running their own copters. "The only real objection to regional programs is the parochial view that department officials simply want their own program. But when we apply hard objective management principles, they choose the regional program for its low cost and effectiveness."

The neighboring counties of Orange and Los Angeles, with the bulk of urban California sprawl, offer contrast. Orange has only recently decided to establish a sheriff patrol. Los Angeles first flew fixed wings in 1924, got its first Bell 47G in 1955, and today has a fleet of public service helicopters, but little in the regional plan.

The Los Angeles "Sheriff's Aero Squadron" at first was largely ceremonial but its usefulness increased with the advent of better airplanes. The squadron now flies four Bell 47G's, four Hughes 300's, three Hughes 500C's, three Sikorsky S58's for rescue, and a cluster of fixed wings for transport and prisoner extradition.

Los Angeles County, 4,074 square miles and 8 million population, extends north to Palmdale on the Mojave Desert, and south to Santa Catalina Island. Bell 47's are stationed fulltime at Malibu Beach, Antelope Valley, and Newhall to service outlying areas. From a Long Beach heliport, Catalina Island is 20 minutes by Hughes 500.

Captain Patrick Devaney, commander of the Aero Bureau, said the unit places police training ahead of piloting

[3] See appendix for explanation of hovering.

skill. (So do most public service agencies.) To fly for this county, a deputy must have a commercial license, helicopter rating and a top medical certificate.

Long Beach, Glendale, Pasadena, South Pasadena, many other cities have helicopter units. The largest in the county, the city of Los Angeles, has 464 square miles of homes, factories, and freeways. Los Angeles police (LAPD) have sixteen helicopters, thirty-five pilots, and twenty-one observers. In addition, city firefighters fly seven copters, and the city's transportation bureau three more.

LAPD does most of its patrol work with Jet Rangers. Three are in the air constantly. Main base and headquarters for this police unit was moved recently from an abandoned airport in Glendale to the roof of a large building in the heart of Los Angeles.

Also high on helicopters, the Los Angeles County fire department owns three Bell 205's and one 204 with large tank capacity for fire work; in addition, a Jet Ranger — painted fire engine red, of course. The 205's can carry 2-1/2 tons of water.

Brush Fires Main Work

Brush fires are the main work of this county unit. Air Officer James Sanchez, who heads the flight operation at a base in Pacoima, said, "The worst happens in the dry summer seasons with high winds. The wind just sucks fires together in greasewood and other brush. We have lost many homes in these fires, you know."

Sanchez described bad flying conditions. "We have hit downdrafts so severe we had to dump the water to avoid hitting the ground. We have to deal with terrain from sea level all the way to 10,000 feet on Mount Baldy."

"But you don't have to have big mountains to have no place to land when your engine quits." (His copters have single engines.)

The unit's ten pilots fly in air as hot as 110 degrees, and "density altitudes of 4,000-8,000 feet about 90 per cent of the time," Sanchez said. They work as many as five fires a day, about 500 a year. They have used over 900,000 gallons of fire retardant in a single year. The unit's thirty-three paramedic squads do rescue work as part of the firefighting routine.

The city of Chicago canceled its police unit. Now the

fire department's sixteen pilots fly on police emergency calls.

The Chicago department created a unique device to fight high rise fires. It consists of a fire hose with nozzle fastened underneath a Jet Ranger. Flying in hover, the pilot can direct a stream of water to any point high or low, on a burning building. It has worked as high as sixty stories, Fire Captain George Tannehill said.

Probably the most persistent voice for public service, Lieutenant Robert Morrison, Huntington Beach, California, Police Department, sees the helicopter in a process of development, "like we did with the automobile fifty years ago. Police authorities resisted the horseless carriage then."

"What they had was not as reliable nor as safe as a horse, and it cost too much. What we have today is hand-me-down helicopters, built for military service, then modified — as best we can — for police work."

"What we got from military surplus," he added, "we had to overhaul completely, sometimes rebuild with new parts, to meet the standards for civilian use. Those hand-me-downs are not cheap."

Other forms of travel have had their share of federal support, he pointed out, with roads and waterways and ground facilities, as well as miles and miles of airport runway, plus untold billions in development funds for the fixed wing.

"It's hard to believe how much money the government spends on new military hardware, airplanes and helicopters —incredibly advanced types — but it goes for warfare, for death and destruction."

"Why," Lieutenant Morrison asked, "can't we use a little federal money to develop helicopters for public service, for civilian use, for keeping the peace and protecting our people? That's constructive."

"Why can't NASA do that, as they've done with satellites, space probes, and the Shuttle?"

"There is plenty of interest around the country in public service but we need better aircraft — safer, more reliable, far cheaper to operate, more suitable to police work — no tail rotor, for instance — than the leftovers we have to use now. There's a big market out there for this helicopter, but the small helicopter industry can't shoulder the cost to develop it."

7.

People Mover

Ground movement by airlines at London's Heathrow and Paris Charles DeGualle Airport almost equals the flight time between the two cities.

— Captain Jock Cameron

Around the world, at least 2,000 helicopters carry passengers, on schedule, to and from offshore oil platforms. Around the world's cities, the total number of helicopters serving passengers on schedule varies from about ninety down to nine. The history of these people-moving services has been quite different.

They are airline twins, working on regular timetables, with the same trappings of check-in and gate call as the fixed wing carriers. Who pays what, and where they fly, set them apart. Big Oil foots the bill for its workers to reach desolate places at sea. Patrons of airlines buy their own tickets, in populous cities and busy airports.

In stark contrast, while oil exploration became the largest, richest endeavor in the rotorcraft world, scheduled city airline transport made a dismal record of fiscal failure. But, as ever in such industry, newcomers keep seeking profits in an alluring gamble.

Two passenger lines have been thriving at Houston, Texas. Others have appeared recently in San Francisco, Chicago, and Atlanta. Of all the city lines, New York and Los Angeles have seen the most success — for a time. Those, too, went to the wall, then were reborn.

New York Airways got off the ground in 1952. It wrote a rocky history in twenty-seven years. It carried millions of passengers on brief runs around the city and to major airports, made a profit one year, and flew various helicopter types. The end came in 1979 following two disastrous accidents.

Before two years passed, two new lines took the field. New York Helicopter (NYH) started in 1981 with forty-eight flights a day between midtown Manhattan and three major airports. Its timetables show 10 minutes from Newark to Manhattan's East 34th Street Heliport. That compares with 1 or 2 hours by auto — and time-saving is the heart of copter service. Other schedules: between Newark and Kennedy airports, 32 minutes (vs. 1½ hours by car); and 6 minutes from 34th Street to LaGuardia Airport (vs. 1 hour by car).

Fares range from $35 to $54 (1984). By interline agreement, fifty-four of the trunk airlines furnish tickets for NYH at a discount or free of charge to their own customers. NYH flies Aerospatiale Dauphins with nine passenger seats and Sikorsky S58T's seating fourteen.

NYH copters touch only one off-airport point but that is enough to suit its promoter owners, Island Helicopter Corporation. Frederic Fine, head of Island, believes it essential for helicopters to operate away from airports. The company has its base in Garden City, Long Island, in a converted bowling alley.

Now Making Profit

In its first year, NYH carried only 50,000 passengers. If that was disappointing, the second and third years did not show the growth its owners wanted. Late in 1983 NYH filed "Chapter 11" bankruptcy and went on operating.

Keith Petsch, vice president, said NYH "turned the corner" and began making a profit in 1984. "We save time for passengers and offer convenience. It is especially good for airline travelers. They are ninety per cent of our clientele." NYH now has 268 flights each day.

This airline's main trouble, Petsch said, is in operating on and around airports. "We have to abide the same air and ground traffic rules as the cumbersome fixed wings — airliners seating up to 300. We wait, we lose the helicopter's advantages. Our line is selling time."

Petsch came to NYH after 27 years with British Airways. He hopes to break the traffic crunch by persuading FAA and airport managements.

Within months of NYH's debut, a second line, Trans

New York, flew a few days, then quit. Pan American Airways set up a competing scheme with a copter shuttle for its own passengers. In the early 1980's, at least 15 new lines took to the air in the USA, although, like New York Helicopter, few got far away from airports.

A more ambitious scheme, Los Angeles Airspur made its debut in 1983, with the new Westland 30 helicopter carrying sixteen passengers. Its routes connected Los Angeles Airport with thirteen stops — nearly all airports — as far as Santa Catalina Island, Oceanside, and San Bernardino. Skeptics wondered how Airspur could succeed where the old Los Angeles Helicopter Airways failed after 25 years.

Then, in late 1984, Evergreen Helicopters — one of the world's largest, with a fleet of 125 copters and some charter airliners — suddenly bought out Airspur. The new owners promptly replaced the Westlands with fixed wings. Evergreen President Del Smith said he saw no reason to fly copters into airports when airplanes would do as well at much lower cost. He reduced the flight schedules and drastically cut costs.

Operating as Evergreen Airspur, Smith grounded the Westland 30's "until we get heliport stops." After a few months of frustration, he shut down the airline.

Meanwhile, opposition to helicopters continues growing in this sprawled city, some of it from airline pilots.

On a more modest scale, SFO, a new line at San Francisco, flies Bell Long Ranger helicopters. It serves airports in the Bay Area, using the same terminals, reservation and ticketing procedures of the major airlines.

A passenger paid $38 for a 4-minute 6-mile zip across the bay from Oakland Airport to San Francisco Airport. He went through the same rigamarole as for the Boeing 747 that brought him 2,000 miles. As the lone rider in the eight-seat copter he was seated facing backwards.

At San Francisco, SFO put him off at TWA's gate, rather than the airline he wanted. That meant walking a mile through the terminal labyrinth. SFO gained him little over going by taxi, and no reason to ride that way again.

Is this to be a service for the rich and the gullible? Perhaps it throws light on why SFO and others do not grow and prosper. The helicopter developed in defiance of accepted rules and customs.

Bell 214ST, one of popular types flying to oil platforms offshore.

Other factors bear on success or failure, such as lack of landing pads, navigation aids, lighting, and other facilities. It is clear that helicopter airlines came into a cold world that relegates them to fixed wing airports — as NYH's Petsch said, among procedures and facilities that reduce their best advantages. Indeed, do they even belong on airports?

Moving Oil Workers

In marked contrast, the oil support lines work away from airports or locate in fixed wing territory for other conveniences.

Helicopters have found their largest business in moving oil workers, especially offshore. This is transport impossible for the fixed wing, and awkward if not unworkable for other means of travel.

Petroleum Helicopters Incorporated, or PHI, has more than 400 helicopters, a fleet exceeding the U. S. Air Force and nine major foreign military services. It is more than the civil fleets of other nations except Canada and USSR. PHI is the largest in oil support — at that only one of twenty-five majors in this field. Its stock is traded on the New York Stock Exchange.

PHI has created systems and facilities which outdo the fixed wing airlines — among which it ranks within the top ten. It was formed in 1949 by Robert L. Suggs, M. M. Bayon, and Frank Lee, to serve oil rigs in the Gulf of Mexico. They still head it.

"Our operations," Suggs said, "cover about 600 miles of coastal waters of the Gulf from Texas to Florida; in New Mexico and Colorado; off the U. S. Atlantic Coast; and in Brazil, Mexico, Zaire, Cabinda, Trinidad, Philippines, and Saudi Arabia." PHI pilots logged more than 5 million flying hours in thirty-seven years.

The Gulf of Mexico, as the world's largest undersea oil field, consists of some 2,400 manned platforms and 220 mobile rigs, plus a number of smaller rigs — the biggest oil community on earth. Almost 1,100 of these have helicopter landing pads. In distance from shore, they range as far out as 150 miles.

This "oil town" has a fluctuating population of 20,000, nearly all male, literally floating. They live ashore and work seven-day shifts with seven days off duty. In some cases (and in other, more distant fields) they work fourteen days with fourteen days off. This adds up to more than 300,000 passengers moved every day by PHI and the other lines.

Copter carriers serve a range of needs, including medical aid. Altogether, nearly 800 helicopters constitute the lifeline for this city on the waves.

Nobody lives fulltime on the water and few workers stay out there more than thirty days at a time. They man the service barges and "oil rigs," enormous flotation devices that ride more or less comfortably with bottoms deep in sea water. Most workers subsist for periods of ten-fourteen days of duty, normally on 12-hour shift. Thus they live within one hour or so from shore by helicopter.

PHI, one of some twenty-five firms serving the Gulf, has six floating bases of its own, for service crews of pilots and mechanics. Its Morgan City, Louisiana, shore base has forty-six helipads, row on row, each with complete tie-down, loading, and fuel lines. Traffic demands three air controllers in the base control tower. Morgan City rivals the main heliport of Fort Rucker, Alabama, the Army's mammoth training center.

Bruce Gillaspie, Morgan City manager, lists PHI's line of shore bases from the west end of Texas to Florida, fifteen in all. "The rigs average 75-80 miles from the Coast — up to 100 miles. Our longest run is 130 miles, though others go to 170 miles; they refuel on the way."

"Remember, when you leave here you fly maybe 30 miles before you hit open Gulf waters. You cross miles of swamp, bayous and salt water flats."

Morgan City is all grubby oil industry, no choice place for residential real estate. Neither is the Gulf climate, hot, muggy, with crashing thunderstorms in the daily fare. Gillaspie lives here, a PHI loner. Of eighty-three pilots on staff, seventy-nine make their homes elsewhere.

"They live all over. One commutes to California for his time off, others as far off as Kentucky, Tennessee, Georgia. Most pilots are married and don't like to move families around in this restless profession. We have quite a turnover."

Gillaspie, at 38, has over 15,000 flying hours, though his job now is largely desk-slavery. In the modest terminal building, the loud speaker frequently spiels off flights like any airline, though the accommodations are plain and functional. The husky passengers, dressed in serviceable clothing, carry small baggage to the scales. They sit around on benches and chairs waiting for gate calls.

Routines here mean flying out to sea, day in day out, through the sweet, soft air of spring and early summer as well as in horrors that stalk the hurricane season. Gillaspie grins when he talks of this as one of the routines. When a big one brews, the helicopters evacuate the great offshore city. In a few hours they can take the manpower and the aircraft to safe haven inland.

Safety Routines Rigorous

PHI has made a good safety record. Discipline is strict because risks can be prodigious in this endeavor. Safety routines for workers and passengers alike are rigorous. All helicopters have pop-out floats for emergency water landings, life rafts, and flotation gear.

"Our policy," said Gillaspie, "is to always manage fuel to reach the beach – plus 30 minutes. We never take off with less. Each new pilot hired gets four or five days of ground school, fifteen hours of flight school – including autorotations onto the water. Regardless of past experience, you start over here."

Training never ends. "Once a year all our junior pilots return for training, including autorotations to full-stop landing. Twin engines go down to a few feet off the waves, then do a power recovery. They don't need to get wet.

"We have few engine failures where any damage is done. We get more trouble from mobile cranes. Sometimes, though, we have to tow a disabled helicopter in rough seas, the tow-line breaks, then there's plenty of damage."

The company seeks mature, self-possessed people with iron nerves, largely family men of senior years. Pilot wages start at $18,000 a year, then progress up to $36,000; they fly 500 to 650 hours a year.

PHI has owned and studied nearly every helicopter made, including several Soviet models. In one of its branches, engineers, pilots, and mechanics test and evaluate new models as they come out.

This company also pioneered offshore navigation. As copters fly below radar range and require many radio channels, PHI formed a subsidiary to develop systems and equipment, such as Flite-Trak, an extended Loran C. It automatically sends out position data, for the moving helicopter, on standard radio channels in this area of heavy traffic. (Loran C is a low-cost electronic navigation aid which provides position data at all altitudes.)

The helicopter fleet had a vital role in oil development, as detailed by Lonnie Nail, president of Tennessee Oil Company. "To one of our platforms, a thirty-six-passenger boat takes 32 hours from port and return, and costs us $22,000. With a Super Puma we can carry the same load the same distance in 2 hours for $5,000. And that says nothing of the accident risk for long hours afloat, and discomfort to employes at sea. The helicopter is our pickup truck."

It took men such as Suggs and his partners, a host of other pioneers in oil support flying, to develop this transportation system of high skills and big money. A few of the large companies, such as Exxon, operate their own copters, some use them on lease, still others contract with PHI or its competitors.

Oil men concede that helicopter services are costly, but they see no other vehicle in the future able to replace it. As one said, "if we didn't have helicopters we'd invent them."

Throughout history the owners, pilots, mechanics, the broad field of users have made this unlikely machine do the yeoman job for oil. They did it their own way, a scheduled system that works well for the customers.

Started in Boston, 1947

The history of people-moving is long and studded with twists and surprises. Tug Gustavson, an industry veteran, insists the first attempt anywhere began in 1947, in Boston. He was a salesman for Sikorsky " . . . and we believed our own sales pitch. We started a service flying from the roof of a downtown parking garage to Logan Airport.

"We had three three-seat S51's and charged $3.50 for the 72-second flight. We made 168 flights a day. Arithmetic was against us. Each trip cost us $20 plus 42 cents' ticket tax. No way we could make any money, yet we were thinking of Los Angeles, Chicago, New York, and San Francisco!

"At that, the service lasted four months. And it was the noise we made, not the losses, that drove us to the wall."

Many others came and went. The English got an early

start, and made more success. They claim British European Airways had the first scheduled line, in 1950, flying S51's between Liverpool, Wrexham, and Cardiff. That and several other tries came to naught, because high costs, low payloads, and no all-weather flying yielded no profits.

Between 1954 and 1967, Belgium's Sabena airline flew S55's, and later, the larger S58. It linked the airport and a city heliport at Brussels with Liege, Maastricht, Cologne, Bonn, Amsterdam, Rotterdam, and Lille. Sabena, too, failed because passengers found it was too often grounded in bad weather.

The sad history of helo-airlines has two noted exceptions, in Greenland and Britain. Steadily since 1964 Greenlandair's S61's have served sixteen towns and villages on the island's west coast. Of the 1,740-mile total route, the longest single run — 195 miles — is barely possible in the few hours of daylight of winter seasons. Over 40,000 passengers a year (equal to Greenland's population) use it.

Of course, helicopter travel in isolated environment beats any alternative. It is similar for British Airways Helicopters' (BAH) line to the Isles of Scilly. There, BAH replaced money-losing fixed wings that flew from Land's End, at the southwest tip of the British main island. Sea fog often grounded the planes, but the copters fly almost free of weather, from a pad in the resort city of Penzance.

One S61 has linked Penzance with Scilly Isles for over 21 years.

Two years after it started, Penzance-Scillies was making profit. No wonder; natives and tourists easily resist the 3-hour boat ride over the shallow channel, the only other way to go. Besides, air fares cost less than twice as much. Best of all, in the copter they can go and return on the same day.

The picturesque isles, noted for balmy weather in spring, summer, and fall, attract tourists from all over gloomy Britain. Their main product is fresh flowers. Scillies is a $36 round trip charge for a pleasant ride from Penzance in the Sikorsky S61N. One rider was heard to say, "It has rather muted noises, don't y'know."

The Scillies line weathered public reaction to its only accident in twenty years, a crash that killed 20 persons. John A. Cameron, retired director of BAH said "the people demanded we continue this line." Population of the isles is only 2,000.

In this classic of airline stories, a single Sikorsky S61 has flown the Scillies route since 1964, with a performance record 97 per cent on time. The 40-mile trip takes 18 minutes at 130 MPH. In summer it flies eighteen round trips a day, six days a week; in summer it usually goes with all thirty-two seats filled. Crews of two pilots and one flight attendant make many round trips without shutting down engines.

BAH has a small staff in Penzance, with four pilots and twelve engineers (mechanics). Routine maintenance of four hours goes on every week night, plus all day Sundays for major repairs, to keep this lone copter working its grueling schedule. All kinks and bugs of the S61 have been covered long since — as tractable as an English tram.

A wall plaque suggests one reason why the government continues this service: Scillies Airport was dedicated by Sir Harold Wilson, retired Prime Minister, who still lives in the isles.

5 Million Carried Safely

In his modest way Cameron said, "Since the BEA Helicopter Experimental Unit was formed in 1947, and until my retirement from British Airways Helicopters in 1981, 5 million people embarked on our helicopters and 5 million disembarked safely. A remarkable record, bearing in mind the state of the art."

He noted that BAH flies between London's Heathrow

and Gatwick airports and serves oil rigs in the North Sea. He conceded the S61 is getting old, "the DC3 of this industry." No other will replace it until the new "third generation" models have gained experience. One possibility, the British-built Westland 30, holds only eighteen passengers.

BAH is the one consistently profitable unit of the Nationalized British Airways. The government in 1984 decided to drop the BAH line between Heathrow and Gatwick[1]; it will close when a new freeway connection between them is completed.

Thus, London has *not* clutched copters to its bosom. The city insists BAH fly only over the Thames river and, being over water, to carry life rafts; in turn that cuts payload and raises costs.

Asked why there are no passenger lines between Paris and London, Cameron said, "Basically, we haven't had helicopters capable of full payload under IFR[2] conditions.

"The S61 payload would be reduced from thirty-two seats (as in the Scillies line) to sixteen — and this is under VFR, without the extra fuel for instrument flight."

Captain Cameron said the forty-four-seat Boeing Vertol 234 "Chinook," largest Western world helicopter, has a range too long for London to Paris, 200 miles. Designed to fly upwards of 700 miles, "economically it's out the window unless we can establish very high passenger fares."

Apparently, BAH has found North Sea oil support profitable, as it operates three 234's out of Aberdeen, Scotland. Flying rates for the offshore services run as high as $9,000 an hour for the 234 Chinook. Among other rates: S61 or Aerospatiale Puma, $2,900; S76, $2,100; Bell 212, $1,800.

To show how Vertol 234 is used, Cameron cited the route between Aberdeen and the "Brent Field," in the North Sea — 320 miles in about 2 hours. Otherwise, passengers go by fixed wing to Sumburgh, Shetland Islands, thence by helicopter, in 5 hours. In bad weather, airplanes cannot land at Sumburgh.

"The Chinook saves 3 hours per man per trip," Cameron pointed out. The oil companies pay employes for travel time to and from port.

A more salty view came from Michael Evans, BAH flight operations director at Aberdeen. "Is the helicopter really acceptable as a public vehicle? The American manu-

[1] London's main airports.

[2] Instrument Flight Rules: flying in bad weather. VFR, Visual Flight Rules: flying only in clear weather.

facturers did not recognize what we need in performance, cost, and the rest of it."

Evans stressed that helicopters have greater exposure to failure than fixed wings, therefore "more subject to emergencies. We should not close our eyes to the deficiencies. I am glad to see they are improving with advanced technology."

This operating pilot, who deals day by day with the hard life of north Scotland, said the North Sea "has pitchforked the helicopter into full use. Bristow Helicopters [a rival] and BAH together carry 1.5 million passengers a year from here."

"We are running a real short-haul airline, day and night, in all weather. It is done by crews who turn around in 30 minutes . . . .It's a horrible job for pilots, unforgiving weather, often sitting on an oil rig in winter winds of 60-70 knots . . . . only six hours of daylight in the Shetlands." Pilots earn up to $36,000 a year.

On a ride with pilots Al Prentice and Manfred Nelson, we moved in calm air over a calm sea one afternoon. From Aberdeen Airport we made the 120-mile run to Shell Oil Company's Stadrill platform in 1 hour, 5 minutes. Except for a rough wooden floor and brocade-covered seats, the S61 interior looked like the well-used airliner it is.

Weather Unpredictable

Prentice had a warning. "You never know what the weather will be like out there, although the forecast is good."

The crew carried heavy blue jackets aboard, but the twenty-six passengers wore bright orange wet-suits. We left the runway pad in short order and long vibrations that died with higher speed. We cruised comfortably at 3,000 feet, 130 MPH. Within minutes most of the riders nodded off to sleep, men returning to duty after a two-week break.

The bleak North Sea and Scottish coast looked prim on a bright afternoon, with patches of sunlight and temperature at 63. Clouds held above 4,000 feet except for scattered ones at 2,000 feet.

Suddenly — it hardly seemed an hour — the platform rose on the horizon, marked by the long mast of its crane. The S61 flew straight in to the pad, hovered briefly at 5 feet, and squatted on a deck covered with rope-matting. Three workmen stood by, ready to remove cargo from under the floor. The passengers filed out, and nine men carrying bags climbed aboard under the slow-turning rotor blades.

In the 7 minutes we spent on the deck we felt no sign of motion. On the horizon, though, the slow-heaving sea was discernible.

The return flight was uneventful, with the lowering sun overheating the cockpit. About 30 miles off the coast, the Aberdeen air radio station asked for a report on "haar, out there, just offshore."

Prentice saw nothing serious to report. "Haar fog," he said, "can get very nasty around here. As the temperature drops, it moves inland. It's a bad hazard for air navigation. Often shuts us down."

BAH began flying North Sea support in 1965. That year it became the first civil operator to win approval for all-weather flying.

In 1983, BAH ditched a Chinook in the North Sea. Flotation devices failed to work properly and the aircraft sank. It was later lifted out and restored to service. Within two weeks, pilots were forced to ditch an S61, which stayed afloat. In both cases all occupants were saved without serious injury.

From his long success in moving people, Captain Cameron sees a big challenge. "We must get the message to the helicopter makers not to underestimate the size of the market or the role they have to play in the development of civil aviation services."

He called for a range of 30- to 240-passengers models as "the right ones" which could "help solve the congestion problems at airports throughout the world. The high cost of operating short-stage fixed wing jets, where not only landing fees but passenger handling, security, and navigation charges have to be paid, have left the airlines with little or no profits.

"In Europe, necessary aircraft taxi time at London Heathrow and Paris Charles de Gaulle airports almost equals the flight time."

In the view of Tug Gustavson, who was star salesman for both Sikorsky and Bell in the earliest days of commercial helicopters, "we are already in the scheduled airline business. The oil industry put us into it in a big way."

Moreover, the lesson is plain in such places as Greenland. "The big market is the underdeveloped world."

Of course, the largest single problem facing the people-movers is their need for facilities, especially heliports. Everywhere, it seems, cities put up opposition, yet without a modest place to land — at least a "helistop" — the helicopter loses its greatest value.

Slowly, the gains come. Helicopters reach every modern hospital and many have landing on their roofs. Private heliports in all major cities now outnumber the fixed wing airports. In California, Texas, New York, Maryland, and Illinois, public heliport construction is increasing. In Ohio, a state program has already developed at least one public heliport in every major city.

There is parallel throughout the past of American transportation, from the days when farmers resisted the "Iron Horse," and later, home owners fought the "Horseless Carriage." In their turn, airplanes had to have airport facilities; these were far larger, and created more disruption than autos and trucks.

Resistance continues to all these in greater or lesser measure. Any freeway or airport proposal today draw stiff opposition. But *progress* has its own life and its own rewards. As the visionaries of the past said, "it will come." Not until fifteen years after World War II did the airlines take the majority of passengers across the oceans. Now the ship lines are hardly more than luxury cruisers. With the rise in traffic went reduced fares. Few can afford to go by ship.

This checkered history strongly suggests that the demand for the helicopter's special advantages will be met, later if not sooner.

8.

How To Handle The Hot Volcano

They never see the helicopter through the camera's eye but it is there.

— Photographer Ralph Perry

With its white snow cap, Mount St. Helens looked like a dish of ice cream on the horizon. For generations it was a familiar landmark visible from 100 miles around, the tall, picture-card peak of the Cascade Mountains in southwest Washington.

Few thought this thing of natural splendor could alarm a wide world, nor draw aircraft as if for battle. But it did, and hordes of helicopters probed every square foot of its great bulk, though not in anger. They arrived from everywhere, it seemed, large and small, civilian, military, police and hired, and not a few tourists.

They served earth scientists, film makers, foresters, sheriffs, and newsmen; river watchers, air controllers, weather specialists, sightseers. Some clients had to see all sides of it, those charged with disaster relief, public safety, air pollution, public health, fire, and many kinds of private interests.

From a Nineteenth Century federal land grant, the Burlington Northern Railroad owned Mount St. Helens. The 9,677-foot peak stood in a section of land surrounded by the Gifford Pinchot National Forest; together with its flanks it edged into four counties. Still, the central fact was not who owned which. What does it mean for mere mortals to *own* a volcano?

This focus of attention had no mines, no resort, no

human abode and, above 6,000 feet elevation, little vegetation and no trees. What good was such a piece of land, rising 9,677 feet into the sky?

Are not mountains eternal, immovable? Native Indians gave this one the lyric name "Loo Wit." Then a great eruption in 1980 brought hoopla that instantly made the volcano a tourist attraction. The authorities built viewpoints and visitor centers for the "Mount St. Helens National Volcanic Monument," duly named by Act of Congress.

Today, inside its crater the live volcano slowly rebuilds; in four years a new dome of lava a half-mile wide reached 850 feet above the floor. The roads and bridges wrecked by the blast have been restored. But within the Monument grounds, the broken trees, debris, and rocks, the volcanic scars will remain as they are, while nature covers all in her own way.

When the volcano first rumbled in March 1980, it hardly puzzled residents, those who worked in its shadow and roamed its flanks. Most ignored warnings — why, it had not budged or peeped in 123 years.

Geologists knew better, and said so. Of the several Cascade peaks, they best knew St. Helens' violent history. Having experienced no eruption, though, they could only guess when or what this explosive mountain might do. At Portland State University, Professor Leonard Palmer noted their prime duty "to know all we can, so we can warn the public. And just as early as we can make it."

Since 1972 the University of Washington had placed instruments about the peak to keep watch on it; these gave the first signals of reawakening. They continue now to tell what the volcano is up to. With other universities and the U. S. Geological Survey they monitor escaping gases, tremors, water flows, bulges, changes of many kinds.

Life went on serene in the gentle Cascade spring as fitful Loo Wit rose from eternal sleep, and reopened her old crater. On a sunny Sunday morning, May 18, 1980 the Monster in the Crater struck without warning. The cataclysm blasted off a cubic mile of the peak. Ash, rocks, pumice blocks, gases and steam shot over 15 miles into the sky. People heard it as far as 400 miles away.

Sound Waves Skipped

Mysteriously, sound waves skipped areas to the south and east. In Yakima, 96 miles east, a Washington Army National Guard troop heard nothing. Nor did pilot Betsy Johnson, at breakfast in Camp Sherman, Oregon, 125 miles south. Nor pilot Dwight Reber at Aurora, Oregon, 75 miles southwest.

The blast originated deep within — 40 miles down, scientists said — from unimaginable power, and continued gushing for four days.

The eruption ravaged whole sections of landscape, largely timber and lumber camps, homes and resorts; it tore out highways, bridges and communications; it set off landslides and floods, and dashed human conceit that "nothing will happen."

It killed thirty-six people, left at least twenty-one missing, and injured hundreds. In the vast mountain wilderness nobody could make an accurate accounting, so those numbers remain estimates.

Because it was larger than expected, the disaster burdened the keepers of order. In turn, they went to helicopters for vital travel. The rotary-winged armada came mostly from civilian sources; that includes reserves of the National Guard and Air Force. They virtually filled the air lanes in heavy weather.

For the sudden demand, air crews had to learn on the job, in extreme conditions without precedent — day, night, high, and low; in dust, ash, heat, weather, winds, strange gases.

Even before the big eruption, police blocked roads because of volcano threat, and air travel became absolute. If you wanted to go high, you went by airplane; if close or low or landing anywhere, by helicopter. As the risk of further eruption heightened, travelers found their only, or safest, way by helicopter.

From helicopters the live and still camera peered into the crater, above the weather, into homes and camps and resorts — before the volcano buried many of them.

Safety rules required *flight plans* for the Red Zone, and only on limited time. Quick trips took earth scientists high on the mountain, to the edge of the crater, and down inside. Landing on slopes became routine. Pilots sidled in to rest a skid on the ground and let people alight on the low side.

Geologists, who often court danger in their field work, gambled with St. Helens, and lost one of their own men on duty. The eruption blew out unexpectedly to the northwest as well as north and east, and hit the campsite of David Johnston, U. S. Geological Survey (USGS) along with those of two other men.

Johnston took his trailer up to Clearwater 2 ridge, about 8 miles northwest of the peak, on duty. He also meant to write a geologist's account of the eruption for the National Geographic Society. By radio he kept in touch with USGS offices in Vancouver, far on the other side of Loo Wit.

Camped near the same ridge were Gerry Martin, 61, radio sentinel for the state's department of emergency services; and Reid Blackburn, photographer for the Vancouver Columbian.

Blackburn, newly married, drove his Volvo there on May 8, for an indefinite stay. He worked overtime on the story for his paper as well as the Geographic. He had two fixed cameras on tripods; a second one, located 14 miles northeast of the peak, he triggered by radio signal, and thus made two different views of the peak at once. On that fateful morning Blackburn was sitting in his car, eating a doughnut.

Meanwhile, pilot Dwight Reber, 36, was approaching to land his Hughes 500C helicopter at Pearson Airpark, Vancouver, Washington. He knew the volcano threat well, but came unaware of the eruption. Until he landed, he heard nothing, saw nothing of Loo Wit hidden in haze.

Reber, a charter pilot for Columbia Helicopters, Aurora, Oregon, knew the peak from flying newsmen, geologists, foresters, and people overcome by curiosity. He had carried Blackburn and Ralph Perry, another Columbian photographer who specialized in aerial work. But on this morning, Reber was to meet a congressman client, up for election.

Plans Changed Suddenly

As Reber landed, Perry rushed over to urge, "Go get Reid. He's up there in it!" He said he would explain to the congressman. Reber took off immediately, meaning to fly over or around the mountain.

A career pilot of keen judgement, veteran of Viet Nam, Reber met a test that day. He locked his mind on rescue of a friend, a young man who had leaned without fear out of his

Columbia Pilot Dwight Reber flying over Mount St. Helens crater.

In extreme cold, geologists study ground inside volcano crater.

To all observers, volcano eruption "felled trees like matchsticks."

helicopter door, at frigid high altitudes, to do his work. He had flown Blackburn and Perry to Clearwater 2, and he knew Johnston too.

As he flew near, an enormous blast plume told why dust and haze thickened. Nothing visible of the peak above the lowest flanks. The column of ash, rocks, and steam towered into the haze, speeding off northeast in the wind.

Poor visibility forced him to reduce speed. His thermometer read outside air at 40 degrees C (94° F). He turned up the ventilating fan in his little cubicle in the alien sky. At 9,000 feet he gave up trying to climb over the ash cloud. In fact, it topped 75,000 feet.

Rounding Loo Wit's west flanks, Reber leveled off 200 feet above the Toutle River; it would lead to Clearwater. It was all ugly mud now. Ash colored everything drab gray-brown.

He flew lower, and slowed to 40 knots. His eyes swept a scene of desolation, mounds of ash, broken logs, rocks. The once pristine Spirit Lake, now a flat expanse of floating rubble, black puddles in matted debris, made it hurt to look. The air felt like a desert sandstorm, sultry, dusty, dark. Steam boiled from the mud, up into black cloud. It reminded him of Viet Nam, and Montana forest fires. No good, he said aloud, time to get out of here.

As he turned, the corner of his eye caught a familiar patch of roadway, a highway and a hillside. He was close. He inched along at 25 knots up over a rise — there in the dimness, Reid's car! Rather, the top of the Volvo. He came to near-hover, less than 40 feet above it, facing the ash-laden breeze.

Ash buried the Volvo up to its blank-black windows, and piled on top. A lonely scene, but peaceful, somehow. Actually, all the windows were blown out, but he could not see that in the murk.

Reber circled again but observed nothing more. No tent. No camera. He dared not land in the ash. Blackburn could not possibly survive this; if he had walked out, he would have called on radio.

He searched the ridge for Johnston's campsite. No sign of trailer or tall antenna, nothing of the geologist. No sign, on the next ridge, of Gerry Martin, just a sea of desolation.

Reber landed at Pearson 45 minutes after he took off. He found the congressman, who stalked off in a huff. The boss later got the man another helicopter; he lost the election anyway.

National Guard Pilot Remembers

One who has vivid memories of the volcano, Warrant Officer Charles J. Nole, 38, pilot of Washington Army National Guard, saw it as stark disaster. He was one of the 105 members of a troop at Fort Lewis, near Tacoma. The Guard Hueys ranged all about that mountain.

On that Sunday of eruption, chance put the troop and its thirty copters at summer encampment in Yakima, 96 miles east of St. Helens. The state had made rescue plans around this troop unit, and it was now able, willing, and ready, but not present.

The troop had no word of eruption until their commander phoned from Fort Lewis to order them back on the double. He spoiled a beautiful spring morning, and the word came too late, Nole said, "to refuel eight of our helos. We had to GO, man, beat that ash cloud rushing on us.

"We got twenty-two of our thirty off by 9:15, and 'red-lined' it at 124 knots. Headed north to Wenatchee to escape the ash, turned west around Mt. Rainier then south to Fort Lewis."

The troop had only Nole and two others on active duty. The rest were civilians who played soldier on weekends. They rather thought of themselves as pilots and airmen who worked part-time at soldiering.

For example, Jess Hagerman flew a Jet Ranger for the Weyerhaeuser Company. His job was to cruise timber adjacent to the federal lands; many acres lay inside the blast zone. Such pro's in the Guard stood willing to do their duty. Now they could only speed back to Fort Lewis as ordered.

Their colonel set them off, as agreed in emergency plans, to transport duty for all vital purposes in the danger zone. That was a rough circle 20 miles in radius from the peak, plus flood areas farther out. That meant more than 300 square miles, for twenty-two helicopters to cover.

The UH1 Hueys had a big empty cabin, which could carry webbed seats for thirteen persons — no rescue gear, meager first aid stuff, limited flight range, but serviceable. Also useful was the five-seat OH58, Army version of the Bell Jet Ranger.

Mud flows, flooding rivers, debris, landslides, melting glacial ice, wrecked highways and bridges all hampered these efforts. Worst was the nasty weather. Cold winds drove rain clouds over the land. As time passed, ground heat dried the ash, and wind filled the air with fine powder dust.

State emergency officials called on military units in three states. Unofficial answers came from dozens of civilian copters as well, but no one kept track of them all. Observers guessed as many as 250 helicopters responded, at various times.

As the mountain continued gushing, geologists found temperatures soared 800 degrees F in the rocks and ash from the volcano. The jet streams carried ash into the stratosphere, took it around the earth to become part of life for months afterward.

The state assigned the Guard copters to sectors of some 30 square miles each. Flight crews had to cope with dangers, new and strange conditions, while scouring the land for the living and the dead.

Always in radio contact, helicopters flew in pairs. While one "Low Bird" skimmed the ground to see better, the other, "High Bird" cruised at 300-500 feet to watch for hazards, to navigate and catch whatever Low Bird might miss. "The buddy system." In case of trouble, one would give instant aid to the other.

They scanned over and over for vehicles, footprints, tire marks, any sign of life. It went slow, in hostile environment. They griped about the bad air, heat, long hours, rough air, everything but the well known risks to themselves.

They carried the living to the nearest shelters or hospitals, and headed right back; the dead they marked with red dye for later retrieval. It was tedious work, hot and dirty, dawn to dusk.

It was disorderly and crude. People complained, but how can you plan for the unknowns of an eruption? The volcano created bad weather and forest fires loaded the air with smoke. Ash laid a gray blanket on the wildly changed terrain. And they could expect new eruption at any time.

"Old people flatly refused to leave," said Nole. "Afraid of our choppers. They had never seen one before so how could they trust a contraption that made such a clatter?"

Cold rain from low clouds on hot ground brought poor visibility and reduced rotor lift. The clouds blanked out the crater, the one thing a whole world clamored to see.

Fuel supplies hampered crews. A Huey burns 80 gallons of Jet-A each hour; they must refuel within 2 hours. The Jet Ranger can go three hours. Pilots constantly plan where and when they will next land for fuel, at an airport or some open spot to meet a fuel truck.

Guard crews cared little for themselves or what they ate, but fussed like mothers over their machines. A gas turbine takes routine care, and the many moving parts of copters are sensitive to abrasive, corrosive ash.

In the time-honored tradition of aeronautics, mechanics gave their best. They flushed the helicopters with water, inside and out, cabin, fuselage, engine. In this episode some put in as much as four hours of maintenance for every flying hour. The days turned upside down for the mechanics, who worked nights so the copters could fly fulltime in daylight.

128 People Saved

The troop flew 684 hours in the first two weeks of the exercise. They saved 128 persons from certain disaster, among the hundreds of others who only needed help. They also retrieved many bodies.

James Scymanky, an Oregon logger, was one of those rescued. He and three other men worked 12 miles northwest of the peak on the morning of eruption. After walking for hours in the hot ash, the four had reached a state of exhaustion.

The loggers saw two helicopters come out of the murky sky but the Guard pilots did not see them. Captain Jess Hagerman said it was "dark as hell down there. We saw their footprints. I don't know why, but footprints stood out in that ash. We were almost on them before we saw anybody."

Flying "Low Boy," Hagerman put one of his skids on the Toutle River mud bank. He kept a ground cushion of air and touched gingerly, then his observer, Randy Santz stepped out. They called but the three men were too weak to walk; one was lying inert on the hot ground. They would have to be carried.

With much difficulty Hagerman landed his OH58 amid the dust, and got out to help. They lifted two men into the OH58 and the "High Boy" Huey picked the third man up at the river.

To keep the two men from going into shock, Santz talked but they were dehydrated after 8 hours without water in the intense heat. Before he fell asleep Scymanky managed to plead for a fourth logger who had wandered off at the river.

They landed on the lawn of St. John's Hospital, at Longview, Washington in 17 minutes. Already crowded with blast victims, St. John's rushed the three loggers to Portland's Emanuel Hospital, which has a special burn center. They flew the 65 miles in 38 minutes in Emanuel's Life Flight. Meanwhile, the two copters returned to find the fourth logger, but saw only footprints leading into a mud flow.

Scymanky survived after a long period of treatment for severe burns on most of his body. One man died in a week, of internal burns, the third man succumbed in early June.

Two months later, another Guard crew searched again. With a dog they traced out footprints and located the fourth man's body snagged in a tree.

The troop had the cooperation of other military units, including the Oregon National Guard, Air Force, regular Army, and Navy. Best equipped and trained for this task was an USAF Reserve Rescue Squadron, from Portland Air Force Base. That squadron lifted out 102 persons; it claimed sixty of these would have lost their lives.

The Coast Guard flew a Sikorsky HH3, from Astoria, Oregon, to join in the rescue work. The HH3 first surveyed navigation hazards in the Columbia River, the Cowlitz, and other Columbia tributaries.

From Whidbey Island a Navy CH46 came in to hover over Spirit Lake and measure temperature and depth of water. Five days after the big blast it was 90 feet deep and 120 degrees F.

On May 22, a helicopter from the troop brought out the body of Reid Blackburn. On his car seat they found two fossil doughnuts, minus bites out of one. Heat had shriveled the body to a fraction of normal weight. The coroner cited death due to asphyxiation.

Weeks later, pilot Reber and photographer Perry returned to the Clearwater ridge, now cool and quiet, though still on hazard alert. They dug out Blackburn's tent and radio control box, and read the final cryptic entries he had logged at the start of eruption, 8:32 a.m., and again at 8:34. His final "4" entry was a hurried scribble.

They found no trace of Gerry Martin, the radio man, nor the geologist, Johnston. David Johnston was last heard on radio, at the time of eruption, "Vancouver, Vancouver, this is it!" At the remote site they dug out Blackburn's second camera. All the films were fogged by heat.

Perry made hundreds of photos for his paper and the National Geographic. Never content, he devised ways to improve the photo image. He fastened cameras on Reber's Hughes 500C, as better than hand-held shooting through the windshield, or through the opening with the door removed. He made novel 360-degree "bug eye" shots straight above the crater.

As a professional, Perry said helicopters in this episode had serious flaws for photographic purposes. "We want a stable platform and freedom to shoot at any shutter speed, any angle, any lighting. You can't open the window on the Hughes. The Jet Ranger's is too small. We could only remove the door and fly in zero cold and cramped positions."

"Still, the helicopter is the best vehicle we have. No other comes near putting us where we want to be, moving slow or stopped. Fixed wings have to go too fast, or have a wing in the way."

The prevailing high rental costs cut missions short or made it too expensive to await better lighting, while the sun or clouds passed.

Perry praised reduced vibration levels with turbine engines. He used shutter speeds as low as 1/60-second. "That's unheard of in the older helicopters with piston engines."

"The layman," he remarked, "has no idea how much we use copters. I see that happening with TV. Viewers never know they're seeing through the camera eye, so they won't believe it is there. Can't tell by blemish on our photos either. I like it that way."

Nature Healing Wounds

Since those hard days nature has done her work. Growth revived in trees and plants and seeds; the deer and elk, coyotes, marmots, raccoons, and little mice found ways to stay alive. Even the workers, residents, and tourists returned. Life goes on.

Eruptive hazards promise to remain for years to come; entry to the Monument "Hazard Zone" will require government permit, and only on limited time. The aircraft remain too (more airplanes now), as they cater to the tourist trade. Helicopters carry more scientists, engineers, foresters, news people, and others on serious business.

In the wake of eruption, fire made hell of green timber. For months, north and east of the mountain, it consumed 61,000 acres and damaged another 90,000 on federal, state and private lands. The total loss was calculated at $134 million.

On one intense fire, in extreme danger of further volcanic blast, firefighters flew in by copter — staying within 20 minutes of safety. Day by day they came and went from Trout Lake, 35 miles east of the fire zone. The volcano burped and rumbled many times, while they labored four months to douse the last spark.

In the course of time the Forest Service, state and private landowners learned to deal with the downed timber. In four years they salvaged 315 million board feet (about 90 percent) of these blown down logs lying outside the limits of the National Monument. Those inside will remain down — a tribute to the Monster in the Crater.

Today the tourists throng there in season; some use the airplane and helicopter rides offered virtually everywhere, for the closest possible view of the crater. Flight operators say their business is steady. Safety restrictions hamper viewing on the ground.

"Nobody," said Bobbie Myers, at the USGS office in Vancouver, "keeps tabs on that air traffic. It varies in seasons but it is substantial, and tourists do more than anyone else.

"On a summer afternoon, as we sit in that crater, I seem to see a helicopter come in every half-hour or so — faces in the window, you know. They come from everywhere, and the seats are filled, at $50 for a 15-20 minute ride."

Myers, a contracting officer, said the Geological Survey monitors water flow, biological studies, gases, tremors, bulges, temperatures, everything about that mountain. The agency has a contract helicopter ready on standby at Cougar, Washington, hardly 15 minutes' flying time from the crater.

The Army Corps of Engineers has a flood control construction project at Spirit Lake. The lake drainage became clogged with debris. In the winter of 1983, when soggy ground made roads impassable, the Corps used Army CH46 heavy lift helicopters to carry five-ton drilling rigs, workers, and tools to the site. The Corps continues to transport employes in small contract copters.

Betsy Johnson's Role

Another who played a major role at Mount St. Helens, Elizabeth K. Johnson, 29, owner of Cascade Commercial Helicopters, Scappoose, Oregon, had a close location, 43 air miles west of the peak. In 1980 she worked seven helicopters, and was herself a member of the U. S. Helicopter Flying Team.

Betsy Johnson was summoned by phone on the morning of the great eruption, at Camp Sherman, far south in the Oregon Cascades. She had made success in a business dominated by men, employed six fulltime pilots (all men) in a staff of twenty-two. As the volcano story developed, newsfolk and official sources clamored service beyond her capacity.

Johnson served the U.S. Geological Survey, Forest Service, and Army Corps of Engineers. The Engineers meant to keep the Columbia River and tributaries open for sea-going ships; for over a year they contended with a huge burden of St. Helen's ash carried by the rivers and streams.

A second large but non-explosive eruption, on May 25, carried backward on winds to the southwest, and loaded filbert orchards with sticky ash; it could break down trees. Betsy flew over them herself to blow off ash for the growers, but claimed "it didn't work well."

Betsy Johnson renamed her company Transwestern Helicopters, but she continues flying from Scappoose, Oregon, Airport. Her fleet has grown to fifteen helicopters.

She believes her best work went to public benefit, taking earth scientists out to study the explosive volcano. This eruption gave them rare opportunity. They placed devices and instruments high and low around the mountain, many with radio telemetering devices.

Today, flying conditions continue to challenge pilots. Warren Fortier, chief pilot for Betsy Johnson, said air traffic still has too many hazards. Winds near the rim of the crater can be violent and tricky.

"We have weather problems and forecasting doesn't reflect events over the crater. I know of two accidents due to high winds. When you sit on the rim you have to be wary

— vortex winds can flip you around just like that. And turbulence is unpredictable."

Fortier said the mountain has cooled somewhat, even the dome in the crater, and the ash problem is down. Worst dangers inside the crater come from dangerous gases and slides down the old crater walls. The abundant sulfur dioxide, for example, forms sulfuric acid.

When a Transwestern helicopter hit a vortex wind in the crater, Fortier recalled, "it rotated as if the tail rotor failed." It rolled over as it touched ground and smashed, but only caused minor injuries. Adverse wind broke up a second copter, parked on the mountain without occupants. Another helicopter carried that one off in a sling load."

Accidents Catch Up

As for the general flying safety record, in more than four years — over 30,000 flying hours in many sizes and types of helicopters — the record shows a number of incidents. None occurred in the first, and worst, ten months of 1980, then a hard landing damaged a Bell Long Ranger from Wisconsin. Its occupants sustained minor injuries.

The two worst accidents occurred within a three-week period in 1983. Both carried tourists, in good weather.

Terry Witham, 37, Toutle, Washington, pilot-owner of a Fairchild Hiller 1100, died in the crash along with four tourist passengers, two couples from out of state. There were no witnesses, and fire burned most of the wreckage. The crash August 13, 1983, took place in woods 8 miles from Toutle. The accident remains under investigation but first inspection showed evidence of mechanical malfunction, and that the pilot had a dangerous heart condition.

On the following September 5, a large copter with pilot and eleven tourists smashed into a rock strewn area northwest of the peak, shortly after takeoff from a viewing site. They were 800 feet above the ground in a Bell 205 owned by Heli-Jet, Eugene, Oregon.

Pilot Harold Kolb reported a "severe shudder and vibration," then the engine failed when he took corrective action. He crash landed in a mudflow, as he was unable to perform a normal autorotation. The impact heavily damaged the floor and seats when the 205 hit the rocks.

The crash injured all twelve occupants, seven of them seriously. One died on October 21, 1983.

Four other copters answered pilot Kolb's distress calls. They brought rescuers, medical aid, and tools to the remote site, and carried nine of the injured to hospitals. The 205 was repaired, to fly again for Heli-Jet.

Some four other persons were injured in three other incidents in the five year period. Meanwhile, fixed wings had a number of crash accidents, including fatalities and injuries.

Betsy Johnson, Jess Hagerman, Dwight Reber, Chuck Nole, Ralph Perry were in a mighty throng of people who found a role on Loo Wit in a time of stress and difficulty. Fixed wings played a part too, but that's another story.

When order reigned once more, it was time to ask how life might have gone without the helicopter. One of its hallmarks is a certain obscurity, to do its work with least possible intrusion on the lives of others. Ralph Perry said "...they never see the helicopter through the camera's eye, but it is there."

Of course, the star of an eruption cannot be an airman, photog, geologist, nor ingenious machine. The years ahead will, we know, write new chapters when Loo Wit, the only star, riles up again, and calls out helicopters or something better in their place. In the way of mountains, that could be a distant age, an unimaginable machine.

National Guard OH58 stirs up volcanic ash.

9.

The Sling Load

I had to hold a piece 42 minutes while they put it together.

— Pilot Bob Brown

"The hooker has a key job. He's gotta be an athlete, move fast through a whole work shift — and woods-wise, ready with quick decisions."

Peter Barendregt points upward. "Right over that hooker's head, there's 200 feet of heavy cable. The hook on this end, that thing alone weighs thirty-five pounds." It dangles from a helicopter.

Barendregt gestures toward a husky young black-beard clad in orange vest and wild-red-white-striped hard hat ("so he's visible"), now connecting a cluster of four logs. The helicopter hovers, impatient, while he hooks the steel cable, or "choker" girding three 40-foot tree-trunk logs. In quick, smooth motion the five-ton load zooms up and away, out of sight over the standing trees yonder.

In seconds, copter, cable, and hook return. Our athletic hooker hops from log to log across the messy landscape. It's a confusion of rocks, broken limbs, debris on the brushy ground.

Barendregt only hints at danger, a man wary but inured to logging hazards after thirty-five years in the woods. Now he is Vice President, Forest Operations, for Columbia Helicopters. With their helicopter sky hook, they call themselves "the Heli-Loggers."

Wiry, athletic and woods-wise himself, Barendregt knows the flying and ground work, what makes good lumber, how to get the most out of helicopters, "yarding methods," (logging

Helicopters lift out logs with least damage to forests.

work). He has put an education in forestry to work on a system that yields over 150 million board feet of lumber in logs in a year. Wood worth $35 million.

He is canny and alert in a skill using machines of brute force. Even the Vertol 107 helicopter fits the macho image. In bulk and shape it seems ungainly on the ground; remove its two three-blade rotors and it looks at home among cranes, "yarders," bulldozers, "skidders," earth movers.

The big Vertol moves in all directions anyway, with grace and ease, in promising endeavor known as *External Load, Sling Load,* or *Heavy Lift.* In the past decade this way of moving big loads has grown. About fifty American firms work at it.

Barendregt, a veteran fighter pilot of World War II, fails to mention self-assurance and sure-footedness as required attributes for his crews. Nobody has to tell them to wear rough clothing, heavy calked boots. They work a 10-hour day in season, a six-day week, and arise early for the best flying in cool dense air.

"The helicopter must go fast. Costs in excess of $1,400 an hour at today's fuel prices. That's twenty bucks a minute. If we go over 3 minutes on a turn we lose money." A "turn" is a round-trip between the pickup and the landing where they assemble logs for trucking to the mill.

Up above, the Vertol hardly stops to snatch the multi-ton load, yet it lifts and sets down logs deftly. To drop a log more than a few feet risks breaking the brittle green wood.

"Sometimes we need two, or three hookers. The helicopter alternates between them. The whole crew is on our radio net with walkie-talkies, in fulltime contact with the pilots." As he speaks, the black-beard hops over to hook another cluster of logs.

Most any method of logging ["lumber jacking" they call it back East] is he-man stuff. When the faller's saw takes the final bite of a tree, the real work has only begun. Now they shear off the limbs, examine, measure, mark, and cut it to "saw log" length as the mill orders.

Moving to that landing means a heavy haul by machine (or draft animals if not too heavy). If it falls on a slope, in muskeg or snow or swamp, or totters on a cliff, the task is tougher.

It takes large helicopters to lift logs, large rotors to make large amounts of lift. They can lug objects of most any dimension or girth, only limited in weight. None today lift over 40,000 pounds.

The external load lifts straight up, is simple to carry out, and works fast — the unique advantages of helicopter lift.

Many Yarding Methods

Men have used animals, waterways, tractors, steam and diesel and other forms of power in "donkey engines," the so-called "skyline" overhead cables, and captive balloons, for moving logs. Except for the helicopter and the balloon (which is limited in wind and weather) such yarding tears up the ground.

The Sky Hook copter's best feature, then, is to transport loads from any outdoor location *externally.* It combines aircraft with derrick, flatbed truck, and barge — a substitute for the ponderous ground apparatus, and works as easily as a yo-yo.

It moves whatever is put on the hook — packages or crates or pallets, bulk materials, wet concrete, firefighters and their gear, wire spools and coils unwinding, bundled Christmas trees, steel towers or sections, cargo from ships, oil drilling platforms, wrecked aircraft, trains or trucks wherever they fall — up and over traffic and other barriers.

Heavy-Lift giants have rotors spanning as much as 65 feet in diameter, 6,000-horse engines, and heavy structure. Two helicopters have lifted in tandem, to handle tonnage over the capacity of one. That scheme rarely is found worth the risk and expense.

Men of the trade revert to the basic question in all heavy lift — for a load worth X dollars what does it cost to deliver?

Barendregt advocates care and planning: where to put the service pad, the landing, the flight routes, how to deal with slopes; it is always better to fly logs down rather than to higher ground; the same for trucking them out.

Mills have sawed logs since the Industrial Revolution put machine power in human reach. They will continue to work with ground machines that cost less to work — as they do now by a large margin. But in certain applications, any less amenable place, the helicopter becomes practical.

Barendregt works with logging tasks that cost less by helicopter, or those that cannot be logged otherwise. The constraints are many. On the ground there is the cost of moving heavy equipment across rough or soft earth, primitive roads, and streams. When the "landing" (where they assemble the logs) is too far out of reach, when road construction creates too much expense, or other obstacles come up, heli-logging wins the job.

Tracing heli-logging history, Barendregt recounts attempts as early as 1958 in Scotland, where a Bell 47D lifted a few hundred pounds, and later in Alaska and California. The USSR tried in 1959. Those all failed for the expense, the dearth of tools and techniques.

"The Russians later reported that helicopter logging *from the stump* without pre-falling, proved impractical. Well, taking the whole tree makes no sense, even when we can handle the weight. Too expensive to move limbs and slash, all that waste, for one log."

Heavy-lift involves a hooker or other key crewman, but the pilots run the show. They have top responsibility and risk the most. They are unseen except for a swatch of face in a bubble cockpit window. Only the two pilots ride the empty machine; they fly light on fuel, too, to put all possible payload on that hook.

To see Columbia Helicopters' pilots you go to their service pad, a small clearing or corner of the landing. Pickups and campers throng the parking lot; crews must live at remote job site for long periods.

It's a mobile force, trucks and vans filled with parts, tools and equipment, and tanks holding 10,000 gallons of jet fuel for the copter engines. Two shifts of mechanics work at least as many man-hours as the pilots do in flying the machines.

The pad also has a small helicopter, a Soloy Hiller 12E seating three; it flies special errands, and ferries woods crewmen to those crags and other spots accessible only to mountain goats.

Pilots fly four-hour shifts, five hours a day maximum, rarely more than 60 minutes at a stretch. They take brief rest breaks while mechanics refuel and check machinery. With feet and hands working the controls, the command pilot works the cable and hook by concentrating his attention directly below the helicopter.

That leaves vital work for the co-pilot, in the right seat. He manages engines, fuel supply, and stays alert for air hazards, weather, and loading. He reads a weight scale on a computer display, at millisecond accuracy. In emergency, either pilot can release the load, the heavy cable, or both.

With its constant slow flying and hovering less than 200 feet above ground, heli-logging incurs some of the worst hazards in heavy industry. "Except maybe steeplejacking is worse," said Barendregt. In nine years Columbia flew its Vertols — over 120,000 hours — without mishap.

Then, on October 14, 1982, with no sign of trouble, a 107 suddenly shed its rotor blades and crashed upside down near Cascade, Idaho. The pilots, Paul R. Rasmussen, 36, and Richard R. Jackson, 33, were killed instantly. A similar accident at Shaver Lake, California, August 20, 1983 had the same result. Killed were pilots Byron A. Harris, 36, and David G. Mann, 34.

In both cases, investigation found undetected flaws in the drive shaft that synchronizes the two rotors. When it failed, the rotors smashed together. The trouble traced to an error in manufacture fifteen years before. Columbia carefully inspected and replaced drive shafts of the type on its entire fleet.

Columbia Founded 1957

President Wes Lematta founded Columbia Helicopters in 1957, with a small Hiller 12, then advanced steadily to the heaviest machines. Columbia's fleet includes one Vertol Model 234 Chinook, largest helicopter in the Western world. It can carry forty-four passengers or fourteen-ton loads in heavy lift.

Lematta continues to favor the Vertol, notably for its absence of tail rotor, which consumes up to fifteen per cent of power but yields no payload. The design also facilitates sideways flight with little regard to the wind. That saves precious seconds in moving loads, without having to jockey and turn for position.

These aircraft were built, he pointed out, "from the ground up for 20- to 25-minute turnaround in passenger traffic. We turn in 2 or 2.5 minutes. We can move quite a tonnage in one 25-minute period. Ten loads in 25 minutes can deliver wood worth as much as $30,000 to the landing."

The Vertol came from military origin, then was modified for civilian passenger service. It presents a curious shape

Aerospatiale Alouette III lifting wreckage of helicopter.

among conventional copters. It has a long tubular cabin and a row of nine (unneeded) windows. Columbia's Vertols came from New York Airways, the Thai government, and U. S. Navy surplus.

To Barendregt it is primarily a "yarder," and he sees it all as a matter of control. "If we can control the 'ship' in wind and weather we can log. That's what counts, control.

"With control, we have worked in 115-degree heat at fairly high elevations, and 40-below in Montana Rockies. Naturally, frigid dense air gives more lift, but our crews were no happier at 40-below that 115-hot. They'd just as soon stay indoors — wind chill in that rotor downwash can be murder."

Columbia has found opportunity in Georgia swamps. In a single month one Vertol logged 2 million board-feet of timber, most of it valuable tupelo (black gum) and cypress. "It was small stuff, no larger than 30 inches diameter, so we cut limbs and took it all out at tree length.

"We only had delays from thunderstorms and heavy rains that hampered ground crews. No trouble with flying there. Nor snakes nor alligators. No real problems except a little muck.

"The bad ones, copperheads, water moccasins, alligators, rattlers go dormant in winter, hardly able to move. That's why we worked in that season. Besides, back home, heavy mountain snow makes it hard to get to the job."

One faller made a wry face about swamp work, but said "it wasn't half bad except you'd get a face full of water when you set a log floating below the surface. The water stayed warm, so we didn't mind being wet all day."

Lematta started in heavy lift with a small construction job in 1961. With a little Hiller he lifted steel girders for some power line towers on the Oregon Coast. Carrying loads as heavy as 800 pounds, he hovered while workers bolted the towers together. He admits it "took some courage."

They did the job in eight hours, versus an estimate of five days by barge and crane — at twice the price. A year later he returned to restore one of the towers after it was toppled by a storm.

Shuttle for 300 Miners

Okanagan Helicopters, Vancouver, British Columbia, another heavy lift specialist, works 15 Sikorsky S61's in a fleet of 137 helicopters. The S61 carries up to 10,000 pounds on the hook.

For a mining firm in upper Alberta, Okanagan did a unique shuttle run for 300 coal miners. For more than two years the S61's carried the workmen in twenty round trips, each day, over a 20-minute, 39-mile route. Despite the winter weather, Okanagan made a performance record of 95 percent success.

The lift saved money and gave the miners the benefits of going home each night. The alternatives were a 4-hour bus ride (at full pay) on rough mountain roads over a distance three times the flight route, or building bunkhouses at the site.

Alaska Helicopters moved several hundred 80- 90-foot complete towers and spools of wire for a power line at Glenallen, Alaska. Loads weighed 7,000-9,000 pounds. On that job, a Vertol and a Bell 204 (civilian Huey) ferried work crews and steel pilings along the 106 miles of line through a winter season.

The job had strict environmental constraints on top of savage weather. Sling loads could not cross the Alyeska Pipeline except where it runs underground; could cross some streams only in certain conditions, never cross certain others. Work halted when Peregrine falcons were seen mating, and when eagles nested in April and May. If they uncovered Indian artifacts only government inspectors could permit the project to continue.

Ground crews could not work in the helicopter downwash more than a few minutes at a time. Wind chill dropped as low as minus-100F degrees. Temperatures ranged between 55-above and 60-below.

In another notable heavy lift, Helikopter Service A/S Oslo, Norway, sling-loaded 500 tons to a high mountain lake for a hydroelectric tunnel job. The lift advanced work on the Abjora project, north of Trondheim, by more than one year.

With a Bell 214B (10,000 pound sling load capacity) Helikopter carried bulldozers, dynamite, tools, supplies, and ninety disassembled barracks buildings, all on a 13-mile route uphill to the lake from a railroad. Loads averaged two to five tons. The single helicopter completed the work in 225 flight hours.

In this plan, crews began immediate work on the tunnel. Then, with men and materials assembled, they started building a road *downhill* from the lake, thereby slashing costs.

General tasks for the sling load make a lengthening list: erecting steel structures, moving bulky air conditioning units to roofs of tall buildings, building breakwaters, moving ore samples, stringing steel and concrete pipe, lifting crab pots and fish catches, ship salvage and towing, making large-scale fertilizer runs for agriculture and forest growth, chemical and forest fire suppression, and oil spill cleanup.

Carson Helicopters, Perkasie, Pennsylvania, installed a 7.5-ton cooling tower on a bank roof in the heart of Philadelphia. The city agreed to block downtown traffic for one hour on a Sunday morning. The S61 put the tower in place in five sections, each lift taking four minutes, for a total of 22 minutes.

In a classic of heavy-lift, Carson's Sikorsky S58T's solved a chronic problem of years' standing — cargo backlog in the port of Jeddah, Saudi Arabia. The jam of ships was so hopeless that arrivals waited six months for unloading, even highest priority cargo waited forty days. It took Carson's S58T fleet a full year to sling load this mass of cargo from ship to shore.

Among other prodigious deeds, Alaska Helicopters towed a loaded hover-barge. For a Sohio Oil Company experiment, its Vertol 107 pulled the vessel over ice-choked seas at Prudhoe Bay. The copter moved the 170-ton craft, with barge engines running, and its 50-ton load, 50 miles in one trip. The copter used a 600-ft. tow line.

Ground speed averaged 10 knots whereas the helicopter normally makes 125 knots. Pilots were forced to crane their necks backward, as the helicopter nosed down by as much as 35 degrees from the horizontal position, against the huge load.

Costs Require Precision

In addition to imagination and knowhow, scheduling and handling with precision on the job are the keys to heavy lift. Large copters cost upwards of $3 million to buy; the largest exceeds $14 million with complete equipment. At those costs they cannot long sit idle.

Future copters will handle greater weights, for the demands of the marketplace. Aeronautical engineers foresee practical limits beyond anything presently in use or design,

Columbia's powerful twin-rotor Vertol towing barge.

but, of course, nobody builds the mammoths until sales promise to defray the cost of development.

As always in aviation, the heavy lift endeavor emphasizes safety, expert care by mechanics, and not least of all, piloting skill. For some insight, read the words of Bob Brown, Evergreen Helicopters, who spoke of sitting 200 feet in the air and holding a four-ton steel object motionless in the wind. Six or seven workmen below trust their lives to pilot and machine, and, said he, "This takes a pilot who is fairly calm."

"When you bring the loads over people's heads you start talking to yourself, not to them. You cannot fly steel and be tense. If you get a death-grip on the cyclic [main control] you might as well go back to the house. You have to have the ability to talk yourself out of being tense.

"One of the first things I do when I go out on a job is gather all the workers and the foremen together and explain just how we work . . . . We'll bring them up to the 'ship,' show them what the sling looks like and how it operates, and tell them what kind of signals we must have.

"They commonly ask, 'Well, how long can you sit there and hold that thing in place while we ding with it?' And I answer, 'As long as I've got the fuel for. We maintain our helicopters in very good shape.'

"If the workmen have a problem with the piece and it doesn't want to fit, we may have to put on a little 'come-along,' drag the corners in, whatever it takes to get this piece to go together — tons of it.

"I had to hold a piece 42 minutes while they put it together — just sit there hovering. Another time, 36 minutes. I held one while they cut it with a torch to force it to fit."

10.

The Juice Will Flow

We run The Lifeline.
— *Bonneville slogan*

Like an arrow we dive straight at an electrical substation. In a split-second, Pilot Bobby Hall averts certain crash into the giant fly-trap. He levels off inches above the massed transformers, wires, and switches.

No, not a near-miss, this is routine power line patrol. In the left seat, Observer John Kohn scans the wires leading from the tangle. Unperturbed, he says, "That line's in the slough there but Dispatch knows. No need to call in." He eyes a thick wire dangling in the water from a broken insulator. That's his job, to find everything that is out of order on the power line system. He marks his notebook.

Hall studies the line too, while working the controls. Easy, as if he is driving a country lane, as if mixing wires with helicopters does not court disaster. As he knows well, flailing rotors make "wire strike" the most common cause of helicopter accident.

Kohn and Hall have teamed on this intensive job six years, day by day, for the Bonneville Power Administration (BPA). Hall's flying trains Kohn's eyes on the power line at close range. These people call wire *conductor* when it carries electrical energy; "wire" is only inert wire. The line carries voltage so great it could fry our helicopter like a bug — voltage so vast as to mean nothing at all to laymen.

This is one of scores of teams across the continent who patrol constantly for a variety of systems which transmit and distribute power, telephone, and TV cable, and pipelines carrying oil, gas, water slurries, and other liquids.

On a network throughout the modern world, above and below the Earth's surface, move life's necessities, the essential, common but invisible stuff of civilization, and all such conduits are served in critical ways by the agile rotorcraft.

Among the many, copters patrol for Consolidated Edison in New York, Public Service of Indiana, United Telephone, Tennessee Valley Authority, General Telephone, PG&E (Pacific Gas and Electric) in California, and the giant Western Area Power Administration which covers fifteen states.

Jet Rangers like those of the Bonneville Power Administration guard the long Alyeska oil pipeline across Alaska. At Glenallen, near Alyeska's mid-point, two Jet Rangers and three Long Rangers work outward, north and south by more than 500 miles. Throughout the year and around the clock they patrol twice each day, and carry maintenance crews to the job site.

Bonneville, oldest and one of the largest power systems in the Western world, has 13,500 circuit miles of two- and three-wire line in five Northwest states — a total of over 43,000 miles of conductor. Patrol assures the delivery of energy vital to some 8 million people in the region.

In season, when the big BPA system generates more energy than the region uses, it goes on inter-tie lines to Canada, and to California, Arizona, and Nevada. The copters patrol these vital lines too.

"Different" Every Day

As Kohn and Hall take up duty one winter morning, their laconic talk reveals little about the day ahead. They schedule themselves under a broad order to cover their whole district six times a year. Their jobs fit within federal civil service routines, for modest pay and benefits.

From base at Portland Airport, they will whiz virtually unseen over a picture-card landscape. Answering our question, Kohn laughs: "Oh no, this is anything but boring. Something different every day." His grin lights a ruddy face.

Dressed casually in light clothing, the two stow leather jackets, caps and gloves in the baggage bin for possible emergency use. Their cabin in the sky is cozy, whereas ground patrols must have parkas, heavy boots, even guns and provisions.

Kohn's briefcase of charts detail the BPA system, every

Bonneville Power patrol helicopters fly 30-feet from wires.

Jet Rangers guard 1,500 mile Alaska Pipe Line around the clock.

pole, crossarm and substation, the minutiae of right-of-way real estate. His territory is one-fourth of the BPA total. The charts cover a land mass the size of the state of Virginia, or five New Jerseys, altogether 10,000 miles of hot conductor.

While Kohn pores over his books, Hall works the ritual of *preflight*. Bobby Hall phones for weather check, fills out forms, and the two nod together a silent "Let's go." In the clinic-clean hangar Hall uses a battery-powered cart under a brown-and-yellow copter to raise it a few inches off the floor. With this he trundles the 2,500-pound five-seat Jet Ranger outside. A mechanic stands alert, hearing with tuned ear.

The gas turbine (jet) engine spins into life, the rotor blades swish-swish over our heads. Hall raises the collective pitch lever for takeoff from the hangar ramp. In seconds we bound up and over the Columbia River, pass a lone tugboat, and head for the substation. Hall speaks a cryptic message to his Dispatcher, the never-sleep know-it-all of the Bonneville empire.

Now, 9 minutes since leaving the office, we enter the bowl of sky swatched with sunlight over foggy river bottom. The broad river divides the states of Oregon and Washington, the factories, docks, roadways, and houses of Portland and Vancouver. Temperature and wind are mild for February, even in this soft climate.

"Looks kinda rough up ahead," Kohn sweeps the scene right and left. "Bet we'll have to peel off in 20 miles." To the west a gray mass of fog hugs the ground.

Five minutes pass. "Let's take it to Longview then, Bob." We turn north to a three-wire line mounted on twin wooden poles with long crossarms between. Hall talks to Dispatch. This line courses the Columbia shore, then climbs high to cross the broad mouth of the Willamette River. We pass above two 400-foot steel towers, then turn squarely north.

Kohn peers through his left-side window, his head moving back and forth as we pass at 50 MPH. We seem close enough to touch the thick wires and crossarms. The poles vary in spacing, six to twelve in each mile.

Our route parallels the Columbia 2 miles to the west. We are on the Oregon side in rough country. The river flows north here for 40 miles, to the twin cities of Kelso and Longview, at the mouth of the Cowlitz. There the Columbia turns west again to the Pacific.

The weather favors us this morning: broken overcast receding in the high sky. Weather trouble is as routine to this team as to any who fly; the worst is fog in the Coast Range, and icing in winter.

Power lines make beelines with tight angular turns. Following the rough ground contour, we ride easy at 15-30 feet above, no more than 50 feet over and beside the wires, dodging high trees and other obstacles in the path. Some of these Douglas firs grow over 250 feet tall; they dwarf the 60-foot poles that were once trees.

Pilot Hall constantly works the controls, his eyes sweep well ahead, then to his instrument panel. The high-whining turbine behind us pitches up and down on his changing power demands.

This patrol combines movement and sound, a live harmony in contrast with inanimate wires. For all its immense voltage, conductor wire does nothing visible — not even a twitch. Movement could mean serious trouble. Flowing electricity makes no sound beyond a low hum, but we cannot hear that.

A River of Wire

The BPA right-of-way flows past as a river of wire. Our windows "snapshot" woods, road, stream, fence, village, isolated farmstead, an occasional car or truck, a patch of sky and cloud. We pass close over a detached, lonely world which looks lonelier still when we go higher to surmount tall steel towers. Some of them stand 525 feet high.

BPA grooms its right-of-way to control trees and unwanted growth. It's a slender lane, 150 feet wide, traceable by tire tracks trailing between and beside the poles. In sudden diversion the trail breaks where the line reaches a river bank or dives over a cliff. We fly on but our eyes can follow the receding track off to one side.

From this cat bird seat we observe how BPA contests nature in this rain forest, with what difficulty every inch of right-of-way must be reached by one mode of travel or another. At certain points only helicopters can enter. And Hall cites a ground track of 57 miles around a curving river; the helicopter crosses in 4 minutes.

The crowding evergreen firs, cedars, and hemlocks menace the conductor. They can whip like knives in a storm. Kohn notes on his charts the over-arching limbs; he will alert the ground crews to come out and eliminate the hazards.

We speed along, 40, 50 MPH, sometimes faster, in sweet comfort. Beside Kohn, Hall notices things too, and occasionly calls out a reminder. Among the worst hazards, they look for leaning trees with shallow roots in the soft damp earth.

"But insulators give us the most trouble." Kohn snorts about "clowns, dam fools with guns," thoughtless sons of bitches who cannot resist those colorful but frail insulators. There are, too, storms, accidents, lightning, acids in the atmosphere, other dangers to insulators, the Achilles' heels of power lines.

The insulators prevent current from reaching the surroundings where it could short out in waste, or injure persons or objects. Engineer Kohn informs us current can "flashover," or jump across a gap; the higher the voltage, the larger the gap. At 115,000 volts, this line can flashover seven inches; at 500,000 volts, 64 inches. BPA uses insulators 3 to 10.5 feet long.

Insulators are made of glass (green or amber) or porcelain (gray or brown). Their polished surfaces shed rain, ice, and snow. They must withstand the pull of wires in the wind. Thus on tension, when hit they explode with a bang in a burst of color.

Of course, there is more. A power line consists of systems and sub-systems engineered to deliver electricity to substations, "ports" where local utility clients take over to deliver to consumers. All facilities are exposed to the hazards of the outdoors.

On its right-of-way BPA marks and lights high towers to warn off aircraft. Even heavy aircraft lose when they hit wires, which may be hidden or shadowed in trees. Pilots have no means to detect wires except to see them. Power lines cause one-sixth of air crashes, and wire strike leads among all helicopter accidents.

Bobby Hall talks the practical view of flying patrol safety. A sudden mechanical failure "might give us four seconds to land. Right where we are. Too low, too slow to go anywhere else."

No, not into wires nor substation tangle, but *close,* or underneath. It has been done, without harm. John Kohn nods assent. Professionals work at safety all the time. They hang their lives on their engine and complex machine, and rely on their mechanics. This patrol has a lower accident rate than the ground crews in trucks. These men know that safety cannot come without planning, nor without a cultivated instinct for averting the faults of men and machines.

What motivates the men of air patrol? Why do they go on with this grubby work, loaded with stress? Hall, who wears a bland face in the air hour upon hour, says, "I like to fly." As indeed he has for twenty-eight years.

With thirty years' experience, Kohn knows the electrical energy, all of BPA's nuts and bolts. "I think this work is important."

Hall observes warning signs atop the poles. They look like regular highway signs. "Danger" signals hazards such as crossing conductor. "Detour" means cut away to right or left and avoid crossing the source of past nuisance complaint — a farm for turkeys, mink, horses, or other animals prone to "spook".

Hall looks for line discrepancies too, so as to anticipate the next split-second move of his observer. He keeps us turned, crab-fashion, so Kohn faces somewhat backward as we pass.

This Pilot No Jock

Hall flies as close as prudence allows. In this exacting task, a pilot of age 48 is no jock. He learned it in the Army. He knows what this machine can do, and what it can't. His pilot career began at age 20. His 15,000 flying hours equal well over 2 million road miles. He has 8,600 hours in the Jet Ranger, eleven years on this job. His face wears the bland expression of one without nerves.

The face of Kohn, a vigorous man of 56, has known weather. His ruddy complexion tells a past of climbing poles, working on "cherry pickers," moving by truck, boat, and tracked vehicle, in mud and ice, on snow shoes. Power lines don't go into places convenient for the linemen. Helicopters do.

The two men have an easy working rapport; they communicate with minimum speech, or none. Bobby Hall darts us up, down, sideways, into turns and climbs and dives, ever striving for a fixed pace, a fixed distance over the changing right-of-way.

Still, this ride is decidedly smooth. We feel little pressure in seat or belts or shoulder straps, less jostling than in highway travel as we sit near the aircraft's center of gravity, and thus feel the least motion. In addition, the spinning rotor cuts through the bumps in the air.

The Jet Ranger has a loud voice — the beat of main blades, a muted racket from the tail rotor, high screams in the slip-stream, the turbine whining behind us at 40,000 RPM, a range of noises.

The radio stays silent, and no excess chatter or small talk on intercom. A few chuckles, some personal stuff. As we dart between tall trees Kohn suggests "we oughtta take the Administrator on that shot there — let him know what it's like!"

The poles flash into view, then past, like marching soldiers. Kohn marks his charts in red or blue. Red denotes something amiss, blue means okay. Red calls out repair groups in trucks; the urgent ones will come by helicopter. Six to eight weeks from now Kohn will do these charts again on another trip. Again, eight weeks later, so as to cover this territory at least every other month. Once a year the ground crews do it at 1/20th the speed.

Ahead looms St. Helens, an Oregon river port. In the foreground sits the Trojan nuclear power plant of Portland General Electric, a private utility. Intense blue-white lights flash on its symmetrical cooling tower, 499 feet high. It stands on bottom land at the river bank. It is the largest American nuclear plant. Four 500,000-volt lines link Trojan to the federal BPA system; we will not inspect those today.

Suddenly, Hall veers off in a fast, climbing right turn. We have been aloft more than an hour. Time for coffee break. The Jet Ranger speeds up-beat at 130 MPH. We zip over the Columbia to Washington in four minutes, head down and land at an interchange of the Interstate 5 Freeway.

Shaking Off the Rigors of the Road

On emerald grass we alight in the city of Woodland, near a popular highway restaurant. We cross the street like tourists come to shake off cramps and fatigue from the rigors of the road.

Over coffee the talk turns to cars, family matters, the price of gas, comments on the route, and rest stops. "We have convenient coffee stops all over the system," Kohn says. "Sometimes we have to land for bad weather or other trouble. We can sit down wherever it doesn't bother people." The patrol will cease such landings on any complaint. Their most novel stop is an abandoned pier at Bandon, on the Oregon coast.

They don't land to loaf but for necessary relief from tension. Whether they stop or not, the work-load remains. "We have to cover a certain amount of line each day or make it up on our own time anyway."

After a rest room stop we are off again. By the clock we resume the patrol in 27 minutes. Now, a higher wind from the left causes Bobby Hall to turn the helicopter at a wider crab angle.

We go into higher, more remote country. The line hangs on tall steel towers. The hills drop off, the line plunges down across the curving Columbia. Intent on work, this pilot and observer seem unaware of the commanding view here.

Four parallel lines, carried on 400-foot steel towers, bisect a mile-long island in midstream. The structures finger the sky from both river banks and both sides of the island. Alders and evergreens flourish in dense patches, and gravel beaches mark the banks.

Thousands of geese float on ponds in the island, white and clean in the sun. They sit undisturbed as we pass. Do these wild things accept the noisy copter-bird as part of this humming power system?

"Ask the birds," Hall grins. Pilots envy the birds' freedom to fly without effort or balky engines, without maps or icing or weather worries. Or government interference. Birds never run out of gas, nor care what it costs.

On the shore opposite, the twin cities of Kelso and Longview straddle the mouth of the Cowlitz River, against a backdrop of green hills and happy white clouds. Here the great arterial Columbia cuts into Coast mountains. Hidden from view 37 miles east is the Mount St. Helens volcano. Ah, scenery — a BPA perquisite.

From our seats, power lines are not without beauty, not without meaning. We know the privilege of the helicopter-eye view. How better take in this scene?

Kohn says they're conscious of it all the time, even when they don't look. The power line adds an indefinable note, this art form in towers and crossarms, insulators and electrical doo-dad shapes of no-nonsense function. And always, the wires are draped in graceful parallel.

Now we follow the four conductor lines in series, four times back and forth we go, through Longview to the edge of Kelso. We fly 5 miles over and back to the substation. On the final crossing, we head steeply down to skim 20 feet above the water toward the towers. The birds remain still, the air smooth, the river glassy, the copter steady as she goes.

Kohn raises a camera to focus. "With this we keep track of bank protection for tower footings." He explains about Columbia floods, how river erosion attacks the island.

The patrol uses lots of photography. Every three years, 100 MPH helicopters cover the entire system with automatic cameras for the benefit of the Line Maintenance Division.

From above, Kelso and Longview look clean and tidy, their riverfront grime softened by distance. The only contrast, volcanic ash rudely stains the clear Columbia waters where the Cowlitz enters.

Climbing back up, we reach the end of our line at the edge of Kelso. There, at a large topheavy tower, BPA current delivers to the Public Utility District. From this point north, patrol belongs to a crew from Seattle.

Close above Longview, we see the intimacy of yards, streets, houses, the orderly urban geometry. Our presence should tell these urbanites that the juice will go on bringing light and heat and the myriad uses of electric power, even the mills and shops, with their jobs — whose products, paper and lumber, bulk large on these docks.

How could Kelso and Longview live without power? We are the sentinel to remind people of power taken for granted, until it fails.

We eye life here closeup - energy delivered, things moved, lights that bless this world, life itself. We know whence it comes. The network of conductor served by Kohn and Hall traces back to dams and other sources such as PGE's Trojan. In minutes we could run from here upstream 100 miles. There the river flow turns the generators of Bonneville Dam. That's where BPA began in 1937.

We turn away at last, a brisk tail wind takes us at 160 MPH back to the hangar, in 14 minutes. We land where we took off, by the hangar ramp and ubiquitous mechanic. It is 12:10, time for lunch from a brown bag at a quiet office desk. On the ground, we find welcome silence with that helicopter shut down.

260 Miles, Three Hours

We have worked 260 miles this morning, about the breadth of Wisconsin. We have flown three hours, at breakneck speed; will do another three hours this afternoon. If we can't do the Oregon Coast line, we will head south, or east over the high Cascades. And tomorrow and the tomorrows after, more of the same. To Kohn, "always something different."

Three other BPA teams fly three other Jet Rangers in the very different geography of the other three quarters of the system. They base at Redmond, Oregon; Spokane and Seattle, Washington. From Redmond they cover largely high semi-desert sagebrush range. The Spokane crews work the similar but higher ground of the Bitterroots and the Rockies in Idaho, Montana, and a corner of Wyoming.

The Seattle crew has Puget Sound and its numerous islands. They fly more hours because helicopters transport people and cargo five times as fast as anything on water. BPA's other two helicopters fly photo, engineering, supervision, and special tasks.

Clearly, power lines make no easy life for workers. Before they had helicopters, though, it was rougher. BPA gray heads know how they did it: "We had no choice. The most hellish weather is what causes the worst to happen, and we were always out in it."

In the decades since, the climate did not change but the hell is mostly gone for these crews who stay anonymous, and who live by the slogan, *"We run The Lifeline."*

Don Ellsworth sees helicopters as flatly essential; they have all but erased service breakdowns. But BPA's chief of maintenance insists it is one element of a combination of means and methods to keep a whole system going.

"We carry power from the thirty dams on the Columbia River System [including the Grand Coulee and McNary dams], so, a transmission line down will darken an entire city.

"For the largest cities we have two and more lines of supply, but we can't sit still with even one out. To be without power is expensive for everybody. Many millions of dollars per hour. We cannot let it happen."

Helicopter patrol costs $15 a mile, he said, compared with $35 on the ground. They don't compare exactly, as surface patrols do tests and repairs as they go. He estimated the copters save, "over $500,000 a year, conservatively." BPA rotorcraft also survey vegetation on the right-of-way; they transport materials, small parts, and workers to remote locations, and carry cargo in external slings.

Like other elements of the system, this patrol evolved over many years. It started in 1947, the dawn of commercial helicopters. Harry Windus, who was longtime head of the aviation branch, stayed with it for 25 years, up to retirement age.

"At the start," he recalled, "they tried a trainer airplane, but in one flight they saw the fixed wing would never do. Unsafe for sure, and it couldn't follow our contours close, go slow or low, stop or hover, and the observer had to ride backwards.

"They tried a helicopter on contract. They had only 1,200 miles of line then, but right away they found so many things wrong, so many ready to pop and not reported — unseen from the ground — the copter was in.

Even so, the primitive helicopters and unskilled pilots made it short of ideal. Somehow, he said, they got by without serious mishap. "The agency concluded it must work its own helicopters with its own people."

The first employe-pilot crashed in 1953, in the only fatal mishap to date, and he was off duty. Ironically, he stuck wires of another utility while girl-watching a beach resort on a sunny day. An autopsy showed he had been drinking.

"After that," said Windus, "we determined to do it right. First, we got control of our shop, for the finest maintenance possible. Then we set out to deliver what our line maintenance people wanted.

"We made up only one of their many tools. We had to win their confidence — especially to know when *not* to fly. We had nothing to lose by being on their team. That worked."

Windus studied their work. "On remote terrain they had to carry materials and supplies, and dig footings, with mules. They carried stuff on their own backs too. We could relieve all that."

Today the BPA crews do construction work on remote terrain — towers, wire stringing, footings, major repairs — with heavy lift helicopters. "Because they simplify the process and do it cheaper and faster. In hours they haul up heavy equipment and tools that would take days and weeks on the ground. And haul it back."

Since 1962, BPA has been using the Bell 206 Jet Ranger, flying new, updated models. Instead of owning, they now lease the aircraft from private sources but continue with their own mechanics and pilots.

Having the first helicopters in the region, BPA often has been called for rescue work by sheriffs and police. Windus recounted one novel escapade, chasing criminals in backwoods country. It came about when the FBI suddenly pounced on him. The agents thought he was aiding bootleggers. "Hell, I was on routine patrol. They only knew about an illegal still — never saw our power line up there!

"When they learned the truth they turned around and demanded help from us, another government agency. So I flew the feds. Dangerous — the moonshiners had rifles and shotguns. They had hounds and geese for watchdogs. So I asked for flight pay." He laughed.

"The bootleggers were real clever. In the end they got away."

Dynamite Topples Towers

BPA air patrol grew with the system. In nearly four decades it made quite a record while sharing in the division's fortunes. Perhaps no single incident stirs BPA memories more than one "J. Hawker."

Suddenly, in 1974, dynamite toppled three towers on a quiet afternoon in the foothills east of Portland. Down went a 500,000-volt line and another of 230,000 volts, major lines of supply to a city of 1 million.

Win Acton, assistant chief of maintenance, was on his way home when BPA's dispatcher alerted him on CB radio, and he sped straight to the helicopter. "It landed us at the blast site in 20 minutes. It was easy to find — big fires were set off by the blast."

The first clues pointed to amateurs, Acton remembers. "The location was too close to a good road. They did not drive to any towers that were hard to reach, never did."

Air and ground patrols swung into action. The six BPA copters flew full-time. Police and FBI went along on air patrols. Next morning, the helicopter patrols found the legs blasted off three more towers, plus other damage in scattered places. No lines went down, but where and when would they next strike?

By letter J. Hawker demanded $1 million ransom, or he would black out the city. His letter belied the clue about amateurs; it was well composed and typed; it detailed an elaborate communications scheme. A later note put the bite on the city's wilderness reservoir as well. He signed, "Liaison, Reorganized Veterans of Viet Nam."

J. Hawker hit eleven more towers and put unexploded charges on others, but he left no clues. No more lines went down, but the weeks-long crisis frazzled BPA nerves.

"Before long," Acton said, "we saw flaws enough, without the clues. These people did not know power lines. It was summer, the lightest power load of the year. One circuit picks up as soon as another goes down. In winter, with a big demand for power, knocking out two lines could mean disaster."

In good time they set a trap. Police bagged an amateur, an unemployed trucker — real name, David Heesch. He had no savvy, no "Reorganized Veterans," no ingenious plan. He earned 22 years in federal prison. His wife's stenographic skills got her 10 years as an accomplice.

Bonneville's helicopters gave yeoman service throughout the episode, with fulltime patrol, surveillance, and transport.

"J. Hawker" taught BPA a lesson in watchfulness. To listen to Acton and cohorts you hear a plaintive note. It was a time of disillusionment to discover a cold world out there. Keepers of The Lifeline were dismayed to find the cold world held more than storms and dumb asses who shoot insulators.

No matter how difficult, they guard 43,500 miles of conductor and all that goes with it in the wide open spaces, in desert, mountain, forest, and waterway.

It took zealous men and women, engineers and pilots and mechanics to develop and fine-tune this endeavor. They remain faceless to the 8 million residents of BPA-Land.

Who does not turn on switches? Keepers of The Lifeline feel gratified when you and I assume the juice will flow. That unfailing flow served to put these words on paper. And it prints and lights the pages for people who read books.

11.

The Wind Steed

I loved hiking but never had the stamina to reach the places we went — in air-conditioned comfort.
— Tom Stimmel

Always, it seems, helicopter people are on the move to distant lands, into strange and exciting deeds. Pilots, mechanics, owners, all live and work with a spirited thing that goes everywhere at will — a *Wind Steed.*

Listen to the voices of experience, of those who have seen strange lands, excitement, fun, and adventures without precedent. They know that with the Wind Steed, lift is where you find it.

The self-assured manner of Bob Ough makes the tale sound ordinary. "It began with CIDA [International Development Agency of Canada], our foreign aid program for the Third World. They contracted with us to do a survey for the Greater African Network."

Could you name a finer purpose than a phone link for the people of Zaire, Uganda, and Tanganyika? These nations emerging from colonial status could take one giant step from the ancient code of drum beat to the direct dial long distance telephone. The undoubted benefits of modern communications, a noble aim to come through three helicopters flown by Ough and fellow Canadian pilots.

What did natives think, those who owned the land, who thrived in their own way before White Men came? Ough has only hints. But he knows White Canadians thought it was a 90-day job. In the 15-man troupe at King City, Ontario — Wilf Seniuk, Terry Jones, Bob Cantin, Bernard Lemieux,

Forrest Eiullett, and Raymond Mercier expected fun, adventure, and good pay as well as world travel.

The task took seven months. Even then they had to rush to finish it, and don't know how they did that well. It was the unique experience that characterizes the helicopter profession.

Africa was in turmoil, as Patrice Lumumba and Joseph Mobutu made headlines in 1974. Westerners knew little of lands whose names changed as they gained native rule.

"Our assignment," Ough recounted, "was precision work, to fix the location and height of microwave towers for long distance telephone lines. Simple task, really. We had three Aerospatiale Gazelles. We'd fly a path in one Gazelle toward the other as a 'slave ship.'

"The slave would start a hover at 250 meters (820 feet) above ground, then inch down *very slow* while we took measurements. Yes, very tedious flying."

The microwave link works like an invisible cable carrying telephone signals, but they must move "line of sight," free of any obstructions, between high towers, from horizon to horizon. It is simple and costs little to operate.

"Beginning about 40 km. [kilometers, 25 miles] away, we took readings of signal strength, then moved up at 10-km. intervals to repeat the process. We'd spend hours mousing around one spot in that terrain - high mountains, volcanos, hot lava, and all.

No Piece of Cake

"Back home, we thought the job was a piece of cake. We didn't know the many sources of interference for a microwave system — so many we could not believe it."

That was not the worst. Zaire's new government was disorganized. "It took a month to get through the red tape. Our problems were due to Mobutu; he brought us in. Maybe we represented what the people didn't like."

They had unimagined troubles. An older resident reminded them that, twenty years before, the natives massacred twenty white priests. The African bush hid lions, leopards, snakes, and tigers. They often saw hyenas, elephants, and crocodiles, and imagined others they never saw. Robbery and burglary were common.

"One night, in Kindu, a modern city, one of our men

was robbed on the street. They threatened him with machetes. Took everything but his clothes." Many times the hapless Canadians were cursed, without understanding a word of the native Swahili.

The Canadian group lived in houses on a lonely old Belgian tea plantation outside of town. One night natives climbed through Ough's bedroom window, with machetes glinting in moonlight; he feigned sleep.

"Two others pretended sleep but their room was locked. They took our valuables, clothes, food, anything they could sell. They burned my briefcase, with all my papers and notes."

The violence came along with hot, muggy, often dry weather — primitive sanitation, dubious food, indifferent supplies. Their maps proved inaccurate or useless. The troupe got dysentery, then the medical care worried them. Three got persistent infections from a doctor's injection needle.

Soldiers took a pilot and surveyor prisoner, one day, as they landed the Gazelle. Informed by other Whites, their fellow pilots went to the rescue but got nowhere. The soldiers gave no reason for the arrest.

"We had to go get the Army commander. Maybe he wanted a ride. We never knew. It really pissed us — seven incidents like that. And we thought we were doing good for these people. We considered canceling out. How could we continue?"

For reasons unknown, the natives did not molest the copters — "fear or mysticism? We never knew. That would have stopped us dead."

"We got nerved up," Ough said. He abandoned caution when a burly guard stopped him from boarding an airliner. "The plane was about to leave, so I pushed him out of the way. At the boarding steps I thumbed my nose at him. Maybe he didn't realize what I meant. But I enjoyed it."

He nurses hopes to return there, "...just gorgeous country I won't ever forget," but sure he would "do things differently."

In Arizona, time stood still when the White Man's flying machine descended some 5,000 feet to Supai Village, on the Grand Canyon floor — the most memorable stunt of Fred Bowen's 40 years in helicopters.

"It was the first of the Bell 204's," Bowen said, "it was night, and like flying into a black hole. We were called to lift a severely injured Indian. It was touch and go with a fog bank up on the high mesa. Really, we saved his life."

Years later, another copter brought a septic tank system to the village, which had never seen a wheeled vehicle nor modern machine, as it is reachable only by 10 miles of switchback trail, a rough one-way path suited to hikers and pack animals. A Bell 214B Big Lifter toted tons of materials, nine 3,000-gallon tanks, 4 miles of concrete piping, and moved a dismantled backhoe to the bottom and returned it at job's end.

The Wind Steed completed the task in four flying hours. The 300 Havasupais prefer this location to any supposedly better life elsewhere. But change has finally edged in with paleface sanitation and flush toilets. The Indians let go of their isolation to entice tourists with green wampum.

Vacation Travel, Too

"It is the best for cross-country flying," says Frank Robinson, of Torrance, California. The annual vacation of a week or two has become a regular thing for Frank and his wife in the small two-place Robinson R22 helicopter.

"Last summer we flew up the West Coast to Vancouver, British Columbia, and on to Port Hardy, at the north tip of Vancouver Island. It was a beautiful trip, with plenty of time for rest and recreation, just under two weeks." They traveled 3,500 miles in 33 flying hours.

Robinson sees it as pure fun, and no one knows this helicopter as well as he, since he is designer, manufacturer and pilot. The R22 covers 300-330 miles at a time, at air speed of 105 MPH, and it runs on regular aviation gas.

The R22 has a tiny baggage space, so what do they do with suitcases? "Oh, there's room. We stuff them under the seats."

Hollywood relies on helicopters for more services than moviegoers know. It makes possible the illusions and fantasies sought for the silver screen, without itself moving in range of the camera lens. The modern copter has such steadiness that expert photographers cannot tell when they use this cameraman's platform for any angle or scene.

Advertising agencies also employ it, as when they perch

the classy new car on the wild desert mesa, or the glacier of a Colorado peak, for that TV commercial. Copter sling load carries the auto to the remote site.

Professionals in flying condemn the film-makers' fondness for linking helicopters to violence. "When was the last one you saw that didn't end up in a fiery crash?" they ask in scorn.

They do those scenes with small scale models, powered by compressed air engines. The models, of course, can perform wilder stunts than the real thing. This has generated a profession of flying models in the film capital.

Not all is bad, however. For one prime example, the 1984 Los Angeles Olympic games, a helicopter served up a grand finale — an enormous flying saucer scintillating high in the night sky.

That saucer flew on a 120-ft. sling from a Bell 214B of Wright Airlift International, Long Beach. They painted the 214 black and flew it without running lights, so spectators could not see it. Two other helicopters, with running lights, flew escort for the 214.

The 4,000-pound saucer had 500 multi-colored lights attached to a 50-foot wheel. The pilot, Bill McMillan, only flew a short distance; he hovered at 700 feet over the coliseum for three minutes. A big effort for a three-minute performance, but it was seen by the multitude there and by hundreds of millions on TV around the world. Even for Hollywood it was rated a stunner.

One wild caper occurred in France, when Monsieur Serge Courtel made the first recorded prison break a la rotorcraft. Police said M. Courtel, although not present, used his wayward genius to mastermind the escape by remote control.

Voila, an aging Alouette II zipped onto a soccer field by the Fleur-Meroglis prison, near Paris, under guards' noses. Two convicts strode up, got in and flew away. Le Surete said the escapees coerced the pilot's help by holding his family hostage. *Alors,* no clues. But the escapees were caught some three months later in Malaga, Spain.

In Poland, 20,000 persons saw an international display of sports flying. The contest at Piotrokow Trybunalski, drew thirty-nine helicopters from Poland, West Germany, Britain, France, USA, and USSR. It stressed timing and precision flying.

In one "slalom" event, pilots flew a rectangular course while carrying a bucket of water suspended underneath on a long rope. The game was to prove skill by moving the bucket between an irregular series of gates at level height, then deposit the bucket on a table. They lost points for running overtime, striking or missing gates, spilling water, and distance off the exact center of the table.

Spilling nary a drop, U. S. Army Chief Warrant Officer George Chrest won the title of Champion Pilot. Others on the team of the Helicopter Club of America were Morton Meng, John Williams, and Army Captain Steven McKee. Scored by the Federation Aeronautique Internationale, world record keepers, the meet saw USA nose out the Russians by two points.

Copter Stories Abound

Stories of flying skill abound throughout the world, notably among military pilots in Viet Nam, but no less among civilians. The freedom of the Wind Steed never ceases to interest them, even the thousand-hour veterans.

Allen Price recalls an antic popular at Bell Helicopter Company. To veteran pilots it may sound daring, but Price claims it is not. "We'd start from 500 feet or more above ground, turn down wind, cut air speed to zero, then give it full left pedal.

"That would kick us around in reverse direction, diving into the wind and gaining speed fast. Actually, it put no stress on the machinery, and was a good evasive maneuver. You could autorotate to a spot that was well behind you."

Across the USA there is hot rivalry in TV Electronic News Gathering, or ENG. It could be an expensive fad, as some critics say, but plainly, ENG is an epic in that mad crush for audience ratings.

From the start, TV ENG has brought hot rivalry among the local stations across the map. Although ENG pervades television USA, helicopters date back into radio days. In 1962 the ABC network made a first live TV coverage of the America's Cup Race, off Newport, Rhode Island. To get the yacht race on the air they put 1,400 pounds of TV gear and paraphernalia in the air.

ABC engineers fought the dearth of electronic gear to meet the weight and space limits of the helicopters of that day. The noisy copter bordered on the indecent, with low payload and high vibration levels. The tiny windows foiled TV's hungry eye.

Electronics never ceases bringing new wonders, as everybody knows. To equip for a flight used to take hours, fussing and testing the broadcasting apparatus, but, no more. Today the whole airborne studio, weighing well under 100 pounds, goes aboard as hand baggage.

Helicopter noise, vibration, and stability have improved; modern "avionics," steadies the copter like a rock, in hover as well as in cruising flight, in any weather. But the broadcasting technology has seen greater improvement. The audience today only sees, nay, expects the camera to poke into the thick of anything. Live, in color.

For instance, a whole microwave transmitter weighs 22 ounces; the old equivalent weighed hundreds of pounds. A "Flight Pac" complete with transmitter, receiver, and power amplifier weighs 26 pounds, and plugs into the copter's 24-volt battery circuit.

Telecasts with such gear satisfy all but the most extreme demands for distance, for fidelity and reliability in the open sky. The local stations, not the networks, do most of the ENG — for which they wince over the costs. Normally they spend upwards of $500,000 on TV devices just to stow in a helicopter that costs almost as much. No plush, yet the investment may total $1 million before they get off the ground.

ENG requires a durable, roomy copter, and they have that fetish about image: what their copter looks like in TV's mix of news, features, and *schmalzy* show biz. Each station wants a shape distinctive from its rivals'.

The stations buy their own helicopters, or lease them at fees over $20,000 a month. Among the popular types for ENG are the Bell Jet Ranger and Long Ranger, Hiller 1100, Hughes 500, Aerospatiale AStar and Gazelle, and Enstrom F28A.

If it was a fad, ENG has become a fixture, and many a station has taken the helicopter to its heart, literally. It perches ready to fly in seconds, from the studio roof or parking lot. Sensitive as they are to public opinion, stations have withstood the hostility that pops out when application for a heliport permit hits city hall. But they cannot do without the Wind Steed.

Oregon Journal Copter First

As "Pilot-Reporter," this author learned the advantages of the Wind Steed with the first news helicopter, flying for the (Portland) Oregon Journal in 1947. It was the first helicopter in executive use, and the first one based on a central city roof.

The Journal's slogan, "Today's News Today" won credence with a Bell 47B, which became the "Newsroom Dragonfly." For more than two years, it showed up best in photography, with freedom for any camera perspective, and delivery in minutes for the photo dark room.

Our associate publisher, Sam Jackson, had great enthusiasm for it and promoted it as fun. He flew it himself, though he had little experience with helicopters. In December 1947 he was killed in a crash. The publisher promptly bought a second 47B and went on with it. We continued flying until our business manager vetoed the expense.

We claimed another Journal first. One morning, as head winds ran us low on gasoline, we landed at a Mobil station and ordered "five of the regular." We drew a crowd of curious in a lonely spot on the Oregon desert. The dumbstruck attendant rushed up to ask, "Ain't you gonna pay me?" This pilot forgot.

In Japan, newspapers make extensive use of helicopters. Many American newspapers today use them routinely through ownership or lease. The competition of televison ENG has stolen the urgency of the print media, however, except on the biggest disaster stories.

Jerry Trimble, owner of Hillsboro Helicopters, landed his R22 Robinson in his home yard one morning. He draped a large sign on it, "It's A Boy!!!" to shout his good news to the world. But an unhappy neighbor signed a police complaint. Trimble was fined, but he took the case to appeal, seeking to overthrow an oppressive city law.

Trimble, by the way, likes nothing better than to go on picnics in the Wind Steed. He seeks the mountain meadow, the sand bar in the river, the tiny islet in the lake. It depends on mood.

Gregory Nutt, a professional ENG pilot, likes to recount his adventures and escapades in his home state of Hawaii, "which is not an easy place to fly because of strong winds and turbulence." He began his far-ranging career in helicopters by trading menial airport work for flight time.

Mark Smothers symbolized the vital role of mechanics in this 1947 photo of Oregon Journal Bell 47B, the first news copter.

BO105 flying upside down over Pennsylvania farm land.

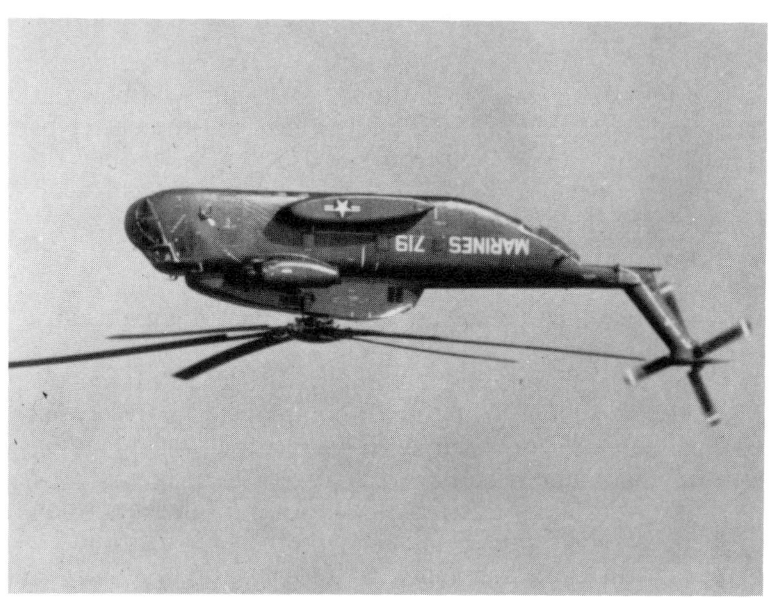

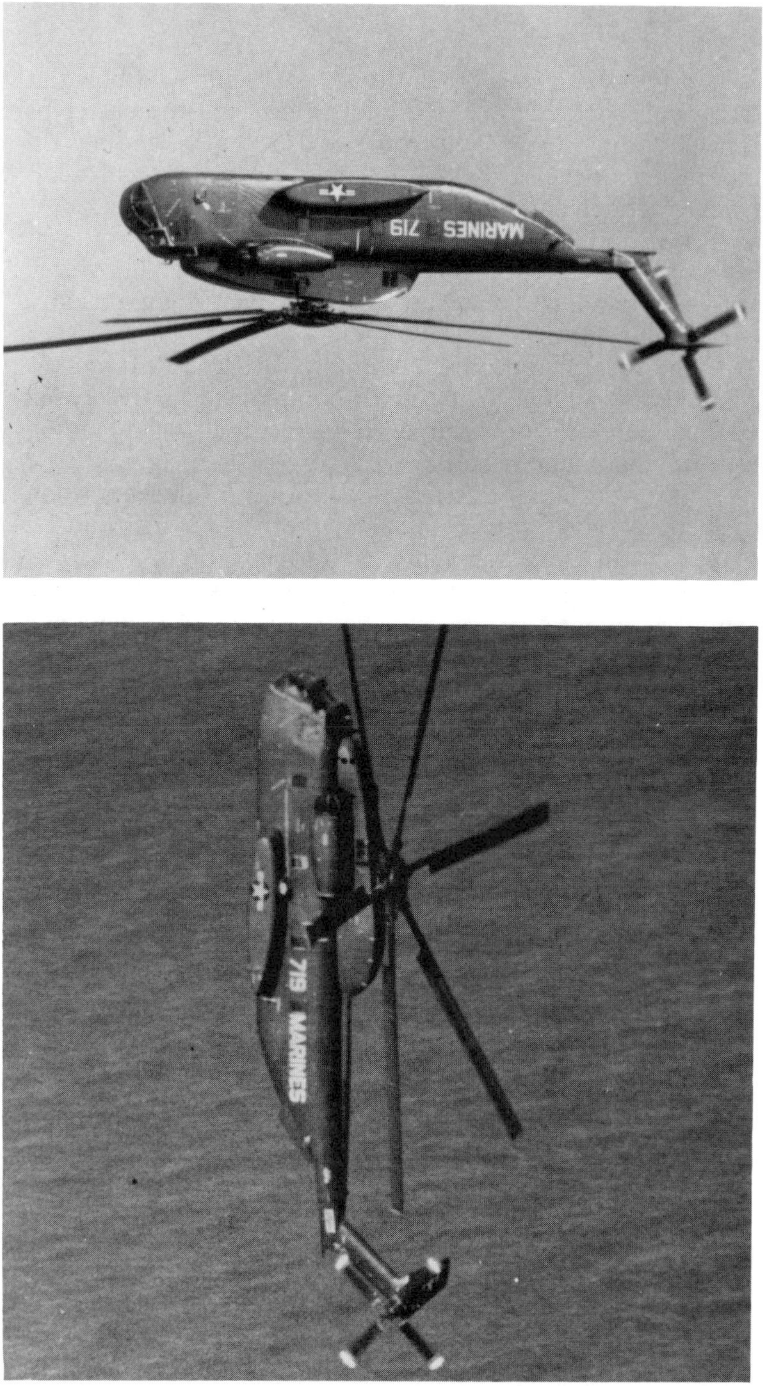

Sikorsky S65 — Yes, helicopters can fly acrobatics.

In the island state he did it all — chasing wild boars in the rough back-country of Molokai, cruising the crater of Haleakala (a rim which tops 10,025 feet), landing on inaccessible beaches below the cliffs of Oahu and Kauai. He did it repeatedly for the tourist trade.

"It's a popular pastime. People pay $75 and up for an hour's ride in a Jet Ranger, but they get breathtaking views. Few people realize how big those Hawaiian mountains are, and how beautiful."

The air can become extremely turbulent. It makes some people airsick, "but nobody complained" in five years. He had his most fun with a dead engine and forced landing on a popular Oahu beach.

"We were in an Enstrom with two girl sightseers when that engine went 'plunk' out over the water. I picked an open spot on the beach and went into autorotation without any problem whatever; I figured it was better to reach the sand than go into the waves. The girls stopped giggling and started screaming.

"We got down OK. I had the only injury, where one girl dug her fingernails into my leg in the excitement. The police came and jawed me about landing on the beach. I asked if they'd like to start my engine. Besides, I didn't endanger anyone."

Nutt and his boss puzzled over how to move the disabled copter. They finally decided to push the 1,500-lb. Enstrom on its skids under a tree. They looped a rope over a limb and strung it up high enough to get a truck under it. Then they hauled it to the hangar.

Nutt sees less fun in such glamor flying than in the obscure work that pilots do. "I like diversification...Guess I'd rather fly forest fires than anything else right now."

Point-To-Point Travel

Helicopters do stunts and set records, although on modest scale compared to the fixed wing. The sales folks of Sikorsky staged a race between Knoxville, Tennessee, and Atlanta, Georgia. In a variation on the ancient tortoise-vs.-hare tale, the S76 Spirit beat the sleek Saberliner jet by many rabbit lengths.

The Spirit outran the Saberliner on straight-line course, 130 miles, carrying two pilots and four passengers. Cruising at 175 MPH, versus the jet's 500, the copter ended five minutes ahead.

Heading straight out from a Knoxville hotel, the S76 beelined direct to Atlanta's World Congress Center. The Saber riders first went by car to an airport, jetted to Atlanta, thence by auto again to the Center. Maybe not a fair contest, but it stressed the reality of point-to-point heli-transit. Consider this score:

	SPIRIT	SABERLINER
Jet-A fuel	79 gals.	145 gals.
Cost	$82.35	$177.94
Miles per gal.	1.64	1.05
Cost per mile	63 cents	$1.16
Ground Costs	0	Cab and limo. fares

The men of Sikorsky denied they meant to twit the fixed wingers, but rather to publicize Nashville's dearth of heliports.

Possibly the strangest copter escapade occurred August 12, 1984 over the Mediterranean, when an AStar refueled a tiny "Quick Silver" ultra-light airplane. That enabled two pilots in the ultra-light to fly 540 miles from Annaba, Morocco, to Monaco. Only the helicopter could fly slow enough for this task.

High piloting skill was required for this unlikely team of fixed wing and rotorcraft. The ultra-light, built for lift, not speed, normally carries a light fuel load and flies for sport. The in-flight fuel transfer took 8.5 minutes while pilots held both aircraft steady on course and altitude.

After landing to refuel its own tanks, the copter was called out again, this time to retrieve "Quick Silver" from the water at dock side. It seems the masts of a bunch of boats fouled the little airplane's landing approach. When the pilot gave it throttle for a go-around, the engine stopped.

Spirits were dampened but nobody got hurt. The AStar pulled the ultra-light out of the water on a 180-ft. sling.

You've heard of jet-set helicoptering to ski spas, so how about a hiking junket in the Canadian Rockies? From Spillamacheen, British Columbia, Arthur Tauck and Hans Gmoser will fly you to farthest wilderness. Their Bell 412 does the worst of the climbing and the distance; the copter ends the pain, sore feet, heavy pack, and long hours of fatigue — and some of the fun. It makes any climb in minutes. You start your hiking labor at whatever height you choose.

Some clients say they have seen meadows or ridges never trod by human foot. "It is scandalous," laughed Tom Stimmel, a news reporter. "I love hiking but never had the stamina to reach the places we went — in air-conditioned comfort."

After you have had your fill, you return to Tauck & Gmoser's "Bugaboo Lodge" for happy hour, dinner, hot tub, and Jacuzzi opulence. So who needs a drafty tent? The hardest part, it turns out, is ground travel; first to Calgary, Alberta, thence by bus to Spillamacheen.

Do-It-Yourself-Kits

There is a small market, at least, for hobbyists in helicopters, and Bernie Schramm, Tempe, Arizona, works to supply it. He makes a "Rotorway" kit consisting of a water-cooled 145 HP engine made by Schramm, plus rotors and necessary parts for do-it-yourself assembly.

With this you get a sleek 90-MPH machine with seats for two. It weighs 1,285 pounds, carries 365 pounds of payload, and runs on auto gas. At $25,000, the price is a fraction of factory-made copters.

The kit includes mechanical and flight training sessions at the Tempe plant, so you can put it together and fly it in the best known safety practices.

But the Rotorway takes a *Zealot First Class.* By Schramm's own estimate, anyone who is handy with tools and willing to work at it every spare waking minute, will take a year or more to complete it. But, look what you get — a high-stepping Wind Steed of your very own.

A wholly new idea was uncorked by Russell Chadwick of Sherwood, Oregon — an ultra-light helicopter. To qualify as *ultra-light* and escape the costly regulation by FAA, it weighs only 253 pounds.

The light Wind Steed will cruise at 50 MPH. It comes complete — no kit — with price tag of about $20,000. Helicopter production is new for Chadwick, a long-time maker of aviation agricultural, cargo, and fire-fighting equipment.

The workhorse copter has found a popular role among coal miners of Appalachia. By FAA count, more than 150 executive models serve this industry. They rove nine states centering on West Virginia.

For example, Ted Hallmark will fly to a mining site and, even in business clothes, run a big machine. Hallmark copters save time in executive travel, deliver supplies and tools, and collect samples for the laboratory. Radios in fifty drag lines and other coaling machines wire together the mining company's operations.

Over the globe, people use the Wind Steed to protect, control, and hunt wild animals. For example, environmentalists in Idaho tried it to save a herd of wild rabbits. The bunnies had been corralled for destruction by irate farmers. These varmints had done the crop owners $55 million in damage in one year.

Orvin Twitchell and Bill Whittom, of Mud Lake, Idaho, said so when they organized the hare war, and they slaughtered 20,000 with clubs. A helicopter sneaked in, buzzed low, and forced the frightened animals to stampede and bolt through the fencing. Nobody owned up to this rescue by Wind Steed.

Deer in New Zealand, however, are a major nuisance. When white settlers imported them, no natural predators existed, and so the herd over-ran the country. Shooting and trapping failed to control them; now government hunters snare the animals by helicopter.

The pilot chases the beast into a clearing where he shoots a net forward to cover it. Then he lands and injects a drug to quiet the animal. He flies the deer, wrapped in the net, to a stockade. Some hunters shoot a dart carrying a tiny radio transmitter so as to track the animal wherever it hides.

Helicopters thus turn nuisance and waste into profit, as the deer are slaughtered for hides, hoofs, and horns, as well as the meat, which has found a big market in Europe.

Business Travel Augmented

The Wind Steed has won place in American commerce, too. Like the fixed wing business jet, it appeals generally to the more affluent, the *upper crust of commerce* largely found in the "Fortune 500." About 1,000 helicopters fly for this purpose, some 10 per cent of the Nation's total civilian fleet. They wear the colors of IBM, Bendix, General Motors, U. S. Steel, Boeing, Rockwell, and many other big names of commerce and industry.

By the cost effectiveness rule they say "if you don't need a helicopter it is expensive. If you need a helicopter and don't have one, it is too expensive."

The corporations trend toward the larger, cushier models such as Bell 222, Sikorsky S76, Aerospatiale Gazelle, Twin Star, and Dauphin, MBB 105 and 117, and Agusta. Down the corporate ladder you find the Long Ranger and Jet Ranger, Hughes 500, Enstrom, even the Robinson.

Within the Los Angeles mega-city, Rockwell ferries employes and clients between plants and airports in three counties. Despite the freeways, or because of them, Los Angeles is the same trap as New York, where average traffic in 100 years has slowed from 10 to 8 MPH.

On the long list, certain corporations decline to talk about their aircraft, claiming "security" reasons. They don't like to flaunt these fancy machines to a public and stockholders who see them as expensive or just plushy. But, in private they cite figures to prove convenience over cars and scheduled airlines.

It is clear that Big Biz is in the air to stay, for practical and prestige reasons. The National Business Aircraft Association asserts that, altogether, some 4,000 aircraft fly for American firms. One quarter of them are helicopters.

Among the copter owners are Timex Corp., Middlebury, Connecticut; Wisconsin Power and Light; Texas Power and Light; Seagrams Distillers, New York; Wheelabrator-Frey, Hampton, New Hampshire; Winn-Dixie Supermarkets, Jacksonville, Florida; Wal-Marts, Bentonville, Arkansas; Allied Corporation, New York City; and Hook Drugs, Indianapolis.

Ted Millar, who heads Westwood Investment Corp., provides an example of executive rotors. His Bell Jet Ranger 206B has a 40-ft. square space reserved in the company parking lot in a quiet section of Portland, Oregon. It's as simple to go in the copter as it is to drive a car.

The company has a major branch at Seattle, about 140 miles north, and major construction sites in many parts of the Pacific Northwest. All of them, said Millar, are within reach to go and return in the normal hours of the business day. They count the Seattle office at "1 hour and 10 minutes from here." They fly in business attire, and use ordinary navigation aids that make piloting painless.

Millar places greatest value on the customers' use. "We take them up to look over prospective locations from close above. They can examine a site from low or high. There's no better way to see how their plant or shop or office will relate to major roads and businesses, everything in an area. Some of our clients insist on it.

"As one of them always says, 'No good to be one or two blocks away from success.' We believe the helicopter is essential for the developer."

Delivering Bank Papers

High interest rates have hurt the purchase of helicopters by business firms but for the banks it has been opposite. Where interest ran up to 21 percent, they found saving of time substantial, for as little as one day, in transferring bank paper.

An Atlanta banker said the price of money got the banks to use helicopters. Under the law all national banks must keep a certain percentage of deposits in reserve accounts at Federal Reserve. As only collected money can be credited to this account, with copters the banks can obtain credit for large amounts of money earlier than other transport would permit.

Atlanta consumes over $300 million a day in checks. If the helicopters can get a portion of this in the bank one day earlier, they are more than paying their way.

The banks fly checks, balance sheets, computer software, and other paper. Where transfer of money between banks would take two or three days, the ubiquitous Wind Steed does it in one day or less. It shuttles between city centers and branches, between suburbs and cities and villages — millions of dollars per month.

Having no suitable landing pads, some banks resort to pickup in flight. The copter flies slow to retrieve a pouch of checks from a hook on a tall pole — reminiscent of the old-time fast express mail snatch by trains. At Tampa, Florida, three copters fly for a group of banks. In 4.5 hours they collect from fifty-six branches within a radius of 75 miles, and deposit at four clearing banks.

Joe Mashman, who recently retired after forty years as pilot and sales representative in Bell Helicopters, had a wealth of experience flying the famous in the United States and abroad. None, however, proved as memorable as Lyndon Baines Johnson, who hired Mashman in 1949 to fly him throughout the state of Texas.

LBJ, then on his first campaign for the U. S. Senate, roamed the wide spaces in a Bell 47D. "After that election," says Mashman, "he loved to credit the helicopter for winning 'by a landslide of 99 votes.' That was his real margin out of a million votes cast."

Mashman's acquaintance continued through Johnson's term as President and in retirement at his Texas ranch. There he kept a Jet Ranger he dubbed the "Johnson City Windmill." Among others, Mashman flew ex-President Truman and the famous Texan, Sam Rayburn, at the ranch. "Truman sat in the co-pilot seat. I think he was 82 then."

"Mr. Johnson invited me and my family for an overnight stay at the White House." Mashman recalls. "A great honor for us. He often called me to his ranch. He liked to consult about helicopter matters. He really loved to fly."

Tragedy of Francis Gary Powers

There is, too, the sad story of Francis Gary Powers. In 1960 he was shot down in a U2 spy plane over Russia. A worse fix is hard to imagine but he survived, stood trial and went to Soviet prison. The USSR held him for three years until they had no further use for him.

Back home, Powers faded into obscurity for a time, then emerged as an ENG pilot in Los Angeles — a career from the highest-flying to the lowest-slowest. On August 1, 1977, while on a news assignment at Santa Barbara, California, he reported low on fuel. After that final message he crashed 10 miles from home. His attempted autorotation failed; he and a cameraman died in the crash.

Powers was known for squeaking by on gas. A probe by the National Transportation Safety Board (NTSB) found only 5 ounces in the tanks and fuel lines of his "Telecopter." NTSB cited this unfortunate 7,000-hour pilot for bad judgment and poor performance.

Between the lines, these stories may be read to imply the Wind Steed demands skill, integrity, persistence, and courage, as is proved over and over by pilots and operators. They illustrate why the copter serves a universe of needs.

Another component of Wind Steed character is the role of skilled mechanics. This is rarely seen or acknowledged outside the shop, but it is never missed by the pilots.

Beyond the U. S. borders, mechanics are known as "engineers." Of course, good mechanics are that; they work as hard as pilots do for their coveted FAA licenses. Not at all your greasy handyman with tool bag, aircraft mechanics wear white coveralls, work in clean environment, study hard in books, and keep their hair combed.

Although the owners do not like it bandied about, helicopters take as much of mechanic-engineers' time as of pilots flying them. In the primitive past that proportion was higher, as much as 10 hours of maintenance to 1 hour of flight on certain models.

Not any more, thanks to better design and materials – and thanks as well to the expert care of the mechanic-engineers. Whatever you call them – mech, technician, or funny name, they own the regard of every man and woman who flies. Simply because nobody can expect to fly safely without their ministrations.

Pilots admit this in grudging, colorful or stony terms, such as "I hang my ass on that mech's ability to fix it right." He sure does, and he can use purple tones when the damn thing goes wrong. Without the mechs' care, he dare not leave the ground. Their care has enabled helicopters to do the prodigious, to perform without flaw in trips around and across the world.

The pilots, the brains at the controls, are the human element aboard. But there is always that somebody else to make it go – from Zaire and the North Sea and Hollywood across this Earth. More than machine and pilot go into the combination. As well, it takes the skilled engineer-mechanic to make the Wind Steed fly.

"Third Generation" Sikorsky S76 Spirit.

12.

Fightin' Helos

Hard to see, hard to hit, hard to down — the attack helicopter fights in the nap of the earth, in the same space as the tank but 2 feet off the ground.

— Richard Tierney

"The danger we face in peacetime, I'm sure, is to get stuck in a mold, complacent, our leadership too doctrinaire. Anything I do with the Army would be dedicated to fighting the next war. And I say the helo will make the tank obsolete on the battlefield."

Strong words from Colonel Robert Fairweather, Fort Rucker, Alabama, a specialist on the attack helicopter — he's a self-styled artillery man who made his career in Army Aviation after graduation from West Point.

"Attack helicopters can hit a tank force from many directions at once, then disperse instantly. I can see the whole saga, how the tank came in and dominated warfare. Now it is the helo, in three-dimensional capability."

The Army strategists demand a helicopter that is hard for an enemy to see or find or seriously damage; one that delivers heavy firepower yet has top performance and speed, a sturdy machine that is easy to maintain or repair anywhere.

Such a tall order would seem a pipe dream to Viet Nam veterans, many of whom curse helicopters as too vulnerable for use on the front lines. Soldiers came home convinced the helicopter is more prone to failure than to enemy fire. It often was, in years past.

But Viet Nam was not all bad. In the final phase of that

war, when TV's "all-seeing eye" could not give full coverage of the fall of Saigon, helicopters performed their largest evacuation of all time. As enemy forces closed in, April 29, 1975, a shuttle of eighty-five helicopters snatched thousands of persons from imminent capture.

For weeks in advance, the Navy and Marine Corps laid plans. Then they committed everything they had in that theatre of war. They named it "Operation Frequent Wind," symbolic of the many copter flights that might be needed.

Planners kept their work secret from the South Viet Namese, and keenly eyed every enemy move at the gates of Saigon. With an innocent public pose, they selected and marked (with luminescent paint) landing spots on fields, roofs, and parking lots about the city. They alerted 7,000 Marines and Sailors, with amphibious forces, ships, airplanes, and attack helicopters to protect the retreat.

At the *GO* signal, at noon on April 29, the flotilla included four attack carriers and ships for two amphibious squadrons, fifteen vessels in all. Into the air, from their floating bases about 75 miles from Saigon, went the big choppers of Marine Squadrons 462, 463, 165, 367, and 369, plus Air Force CH 53's aboard the carrier Midway. This mass force had no less than forty-four CH53s and twenty-seven CH46s plus six UH1Es and eight AH1Js.

Early that morning, enemy action spoiled plans for any evacuation by transport airplane from Tan Son Nhut airport. The runways of Saigon's big base were blocked by disabled aircraft and trucks.

"Helicopter operations were continuous from first light," the evacuation commander, Marine Major General Richard E. Carey, reported, "until the operation was concluded at 8:35 the following morning. The statistics compiled were record-breaking...." Twelve CH53's brought in Marine landing specialists, then ferried 679 persons to the ships in the first wave. In the exercise, they made 682 sorties altogether.

The shuttle went on at fast pace through the night, in spite of bad weather, city smog, smoke from burning buildings, and heavy enemy fire with small arms, artillery, and rockets. The Marines also dealt with communications breakdowns as well as confusion at the landing zones, where thousands of desperate people assembled.

They flew 360 sorties at night. Each helo flew at least 1.5 hours on the round trip between the ships and landing zones. Most flights were conducted on instrument conditions. The helicopters logged a total of 560 hours, day and night, for an average of 13 hours each; the longest, by Captain Gerry Berry of Squadron HMM 165, put in 18.3 hours; he and his crew worked non-stop through it all.

Two Pilots Killed

The harried operation lost only two aircraft and two Marine pilots. Two were killed when their search and rescue CH46 crashed into the sea. In a second crash, General Carey said, an AH1J ran out of fuel when a civilian copter took up the deck space where it was supposed to land. Its crew was rescued unhurt by a whaleboat.

At the American embassy compound, the Marines planned to fly, at most, a few hundred persons off the roof and the very small parking lot. But 2,000 men, women and children ganged up there. Marines scheduled one CH53 and one CH46 every 10 minutes; even then, some were left behind. The ambassador himself left at 4:58 a.m.

The last of the security forces, under Major Jim Kean, withdrew into the Embassy and barricaded the doors. Then, General Carey said, "they moved up through the building until they occupied only the top floor from which they had access to the roof landing zone. After dodging small arms fire and finding it necessary to employ riot control methods against people attempting to storm the roof, Major Kean stepped aboard the last helicopter for liftoff at 7:53 a.m."

The Marines finally shut down at Tan Son Nhut airport when enemy fire grew too hot. At 1 a.m. they abandoned the place and blew up the buildings. By that time they had rescued 4,870 persons from that point.

In all, "Frequent Wind" extricated 6,968 persons to safety on the waiting ships. The total included 1,373 American citizens. As Carey said, this prodigious feat didn't just happen; it was planned.

Still, helicopters have their faults and their scoffers, and all is not easy among various factions: those who plan to fight that next war, those who will fight it, and those who pass judgment on those who do the fighting. Authoritative sources count more than 26,000 helicopters in the world's fighting forces.

Significantly, American forces have no more seaplanes. The helo has replaced them. Warfare is changing, as it ever does, and change today means more helicopters for Army and Marine forces, at least — more rotorcraft for fighting, not only transport and rescue. The age of copter dogfights lies ahead — the attackers fighting other attackers, at extremely low altitudes.

For this future, the Army has two advanced new models, the AH64A Apache attack helicopter, and the UH60A Blackhawk. In addition, the Blackhawk, in different versions of one basic airframe, will go into service for Marines, Navy, and Air Force.

Both new models emerged from long development process, and though they were produced to fight and destroy enemies, they will bring advancements, in time, to civilian models.

American armed forces own more "choppers" than airplanes. The Army has some 10,770 helos, Navy 735, Marines 780, Air Force 275, and Coast Guard 178, for a total over 12,750 of many types and sizes. This compares with some 12,500 fixed wings. The Army's 11,000 aircraft make it the world's largest air arm.

Copter Speed Important

"To fight and win outnumbered," an Army manual says, "forces must move rapidly to concentrate at critical times and places so that a reasonable force ratio exists at the beginning of the battle." Meaning, "git thar fustest with the mostest."

Fighters now face guns and missiles ten times as effective as in World War II; they can hit targets accurately within 2 miles. But "attack helicopter units can fight and survive on the battlefield in spite of the increased range and lethality of modern weapons."

As modern armies are completely mechanized, helicopters add a third dimension to battlefield mobility. Teamed with fast ground vehicles, they concentrate combat power on targets.

The U. S. Army claims the Hughes Apache, a ruggedly built two-man, twin-turbine, killer copter will operate night, day, and in all weather conditions. It bristles with 16 Hellfire missiles, 76 rockets, or 1,200 rounds of 30 mm. cannon, or a combination of those. One Apache can destroy two tank platoons, so the Army says. Or it can destroy light armor and troops, or support infantry attacks.

Not large for a helicopter, the 48-ft. AH64A has a 48-ft., 4-blade main rotor, height of 17 feet, and stub wings measuring 17 feet tip to tip, which carry fuel and armament. The Apache gunner sits up front, the pilot just behind and above him. Either crewman can fly the craft or run its armament. Between them, armor plate prevents disabling both crewmen with one shot.

With an array of electronics and lasers, the Apache pilot can find and "see" a distant target day or night, in haze or smoke, or light rain. The gunner can aim laser-guided missiles directly, or in teamwork with others on the ground or aloft. The (infrared) night vision sensor puts data on a monocle in front of the pilot's eye, and that instrument is made a part of his helmet and display system. In addition, the thing displays airspeed, altitude, and other flight information — elaborate aids to simplify the flying task.

In performance, the sponsors say the Apache is tops. It will go as fast as 245 MPH, up high or down in the weeds. In the so-called "nap of the earth," the attack copter evades detection on radar or other instrument, thus working surprise on the enemy. On the other hand, it can accelerate, turn, pop up or down in flight faster than any other helo made — to take punishment as well as dish it out, like a fighter airplane.

Flying low means hugging the ground, day, night, in the worst visibility. By concealing itself among trees, hills, gullies, tall corn, brush, and buildings, says the Army, "it will be extremely difficult for an enemy to successfully engage the Apache with air defense weapons." Its dimensions give it a small head-on profile, making it that much harder to see and difficult to hit, directly in front or behind.

The Apache's main rotor has no telltale "flicker," to impede detection by radar. Tail rotor noise is low by design, and special devices reduce the engine exhaust heat to foil detection.

Even when detected, this post-Viet Nam model is built to survive the rigors of battle — light machine gun fire, incendiaries, and explosives. Its gear boxes will withstand a hit — they use grease instead of oil, and will continue working for a full hour without failure. Armor plating protects rotor blades, transmissions, engines, and other vital points. The seats and the sturdy landing gear protect the pilots and the structure in hard crash landings.

The AH64 was designed for simple maintenance pro-

High-speed Army Apache attack helicopter shooting Hellfire rocket.

Hughes AH64 Apache in demonstration trials over desert.

Sikorsky H60 Blackhawk, new workhorse of U. S. forces.

U. S. Coast Guard's all-purpose Aerospatiale HH65N Dauphin.

cedures, with minimum lubrication, fault detectors and locators, quick release pins to permit changing rotor blades in the field, and built-in work platforms for engine service.

There is more to this replacement for the chopper of past wars, of course. Its good news arrived with bad. When delivered at last, several years late, the price tag for the basic machine had skidded from an original $3 million to $7 million each, one of the costliest helos ever built. With full equipment the Apache tops $14 million.

The Army staged a ninety-day testing period in desert conditions to prove the AH64. In an unusual 410-hour "fly-off" kind of contest with critics and weapons from other services, the Apache is said to have done better than expected on all counts.

Apache History Long

The Apache history dates back some thirty years, as an outgrowth of the Korean War. Lockheed Aircraft produced the first one, an XH56 Cheyenne, but the Army canceled it in favor of a modified Bell Huey Cobra. The Cobra had the attributes and drawbacks of the basic Huey. The Army ordered the wholly new attack chopper in 1973. Hughes began the AH64 development program in 1977.

Already the Army is planning improvements, as AH64 deliveries begin to flow. One change will put durable rotor blades of composite materials in place of the metal and fiberglass of this model. Another will replace the cyclic collective and pedal controls with a "fly by wire" electronic system. The pilot will have a small "side arm" lever instead — edging into the Buck Rogers age.

Army leadership sees a symbol in this first helicopter designed to kill tanks, a means for Army aviation to come of age, "a helicopter to live and survive on the battlefield."

Says Richard Tierney, civilian editor of Army Aviation Digest, "Hard to see, hard to hit, hard to down — it flies and fights in the nap of the earth, in the same space as the tanks, but 2 feet off the ground. It takes advantage of the same cover the tank has, but many times faster.

"That's far better than what we did in Viet Nam, go up and over ground fire, then down to fight."

Virtually the same rugged characteristics, the same performance feature of the UH60 Blackhawk, but its roomy cabin was designed for the Army foot sloggers. The

Blackhawk comes close to full replacement for the ubiquitous trucks that Soldiers ride; in fact, it's a carrier for the basic element of the Army, the squad of eleven men and their gear.

It too has high performance, cruising speed of 175 MPH, high crashworthiness, survivability, and simple field maintenance. Where older helos required maintenance at 100 hours, this one goes 500 hours between checks. It has a crew of three.

Made by Sikorsky, the Blackhawk has two engines similar to those in the Apache, the 1,500 HP General Electric T700. Mechanics can replace the engine in the field, with ordinary tools, in half an hour, without safety wire or other repair nuisances.

Blackhawk comes equipped with an APU (auxiliary power unit) for runup checks, starting, and emergency use; automatic flight control system and a full array for instrument flight. It costs more than $6 million.

The Navy's SH60B "Seahawk" has major differences, as it will base aboard ship for anti-submarine warfare. Its main and tail rotors fold away for ship storage. It has a hydraulic hoist for rescue work. A hydraulic winch-cable device will hold constant tension on the Seahawk to make it easier to land on a heaving deck. The cable then pulls it into the ship hangar.

An Airborne Tactical Officer seated in the rear — not the pilot — commands the helicopter in his role of sensing and tracking enemy submarines. It works as an extension of the destroyer, cruiser, or frigate, and carries two torpedoes or depth charges.

Still another version, the HH60D "Nighthawk" does rescue for the Air Force; it is loaded with instruments and special devices for finding and rescuing airmen at night and in bad weather. It also has external fuel tanks for extended range. The USAF hints at "special operations" as well.

The Apache and the Hawks are two of the newest, most promising models in the military stable. There are others, of course, largely improved and modernized older ones such as the OH58 Jet Ranger. In the main, the changes improve performance and, as much as possible, reduce upkeep and repair. Some of the Viet Nam copters took as much as 15 hours of mechanics' time for every flying hour.

Coast Guard Goes Own Way

The Coast Guard, one of the earliest American forces to enter the helicopter world, has gone its own way with an order to replace its entire ninety-helicopter fleet. After protracted deliberation (and no little political heat), it took the unusual step of ordering a foreign helicopter, the French Dauphin 2.

Designated HH65N, the Dauphin is adapted to the Coast Guard's role in search and rescue. With a 39-ft. four-blade main rotor, it has two 850-HP (American) Lycoming engines, and cruises at 170 MPH with a crew of three and a passenger capacity of twelve.

The Dauphin was designed and developed by the French, and is assembled at the Aerospatiale Helicopter Corporation at Grand Prairie, Texas. Its engine and more than half of its parts and equipment are American-made.

Foreign nations have been busy with military helicopters too, notably France and USSR. The latter keep their work under wraps, as always, but they make the world's largest and heaviest models.

The Germans and French today have a military partnership born of the cooperative spirit of the Common Market. Together they plan to develop and produce an attack helicopter with certain features for each country's own specification. The British and Italians have a joint product in this field, as well as military copters of their own.

In a world of big munitions budgets, big manufacturers happily take the big military contracts with guaranteed profits. Such was the pattern for aviation in World War II and all wars since. And of course, that was the source of numerous aircraft which later went into civil use. So much is history — a very long history that goes to the roots of rotary wing flight.

Two new ideas show promise on this front, the XV15 Tilt-Rotor of Bell Helicopter Textron, and Sikorsky's "ABC." Both involve radically different design ideas.

The XV15 results from more than thirty-five years of research to produce a mutation of helicopter and fixed wing. This one received more research and development aid from NASA than that agency put into any other helicopter program.

The XV15 consists of twin 25-foot rotors mounted side by side — or the fixed wing buff may describe it as twin-engine airplane with 25-foot propellers on a stubby wing. In takeoff and landing, the rotors work like a helicopter; in forward flight, they tilt down into propeller position. This design makes possible speeds above 350 MPH.

Bell has teamed with Boeing Vertol company to tout the advantages of the Tilt Rotor for military uses. Of course, it has potential for civilian use as well.

The ABC or Advancing Blade Concept of Sikorsky represents the first attempt in recent years to produce a twin rotor coaxial design (two rotors mounted one above the other on a common shaft). The company has pushed this development; under Army-Navy sponsorship it became the XH59.

The ABC has a large double-fin airplane tail, and two jet engines mounted on the fuselage for added thrust. Its two hingeless rotors are devised to reduce the drag on one side of each rotor, to enable higher speeds and great climbing performance, as well as a smaller, more compact frame.

This design eliminates the tail rotor and puts more power in the lifting rotor system. The ABC has topped 300 MPH in tests. Sikorsky has no contract for production, but is continuing development work.

The generous sums of money going into military development today will, no doubt, strongly impact the future of the civilian helicopter — for good or ill. The fightin' helos are finding an ever larger role with armed forces everywhere.

MARINES

13.

Meeting The Challenge

Man wants to fly like the birds, not like a bat out of hell.

— Lawrence D. Bell

In 1863 a Frenchman, Ponton d'Amecourt, flew a tiny helicopter model. Steam power turned its two rotors in opposite directions, one above the other on one shaft — the first so-called "coaxial" design. d'Amecourt wrote a detailed description of his remarkably prophetic ideas.

He had no adequate power plant for his invention, yet this visionary dared to predict, "The flight of the helicopter will be possible tomorrow." Tomorrow did not arrive until the 1930's when other men built and flew full-size copters (with gas engines) based on principles that he and others discovered.

d'Amecourt was one of many inventors. Helicopter history, in fact, began in antiquity. It was 1483 when the Italian Leonardo da Vinci, designed one with broad wings. He made no more than drawings of what he called an "aerial screw." In 1843, George Cayley, an Englishman, designed both helicopters and fixed wings. Such were the beginnings, and (as with airplanes) helicopter development depended on the progress other men made with the engines to power them.

Now, near the close of the 20th Century, we have engines of great power and reliability, and helicopter flight is virtually automatic. Take off in a new "Third Generation" model. With small hand motions, the pilot sets course. As the altimeter needle passes 300 feet, eight passengers relax in plushy seats while this machine flies itself.

Pilots push a series of buttons to program our flight of 216 miles, just over one hour in this wind. No need to work controls until down within 50 feet of the ground, ready to land.

The electronic "brains" in our machine not only do it all, they tell us in plain words how far we are, wind speed and direction, how fast we travel over the ground — they even inform us of errors, if they make any.

The "How Goes It" instrument constantly reads off the how, when, and what of the route ordered; go to 1,400 feet, follow headings, turn for shifting winds and temperatures; run anti-icing systems as needed, and keep the cabin at even temperature, warm or cool. Sure, it will make coffee and tea, if you like.

Surrounding the cockpit of muted lighting, a complex of gauges, handles, and switches run the electronic devices that control this machine. In front of pilot positions, the cyclic control knobs have five buttons; at left, the collective levers have more such fingertip controls. The sleek exterior covers a mass of electric, hydraulic, pneumatic, and mechanical complexity.

Bet on it; this machine has integrity. Slightly less than human, if not more reliable, it will miss nothing on this passage — except watch for air traffic or other hazards. That's what pilots do, though it isn't much of a job to make radio calls and use our eyes.

This classic of sophistication could be a Sikorsky S76, Bell 222 or Long Ranger, Aerospatiale Dauphin 2, MBB-Kawasaki 117, Agusta 109, Westland 30, Boeing Vertol 234, any of a dozen different new types. They are the most advanced in modern technology, in aerodynamics, gas turbine (jet) power, structures and composite ("plastic") materials, and avionic systems.

Still, in most ways the helicopter appears little changed after the early production models, with wide rotors, skids or wheels, a long tail, and a functional, lumpy shape. On the ground, at rest, it won't compare in looks with the neat fixed wing jet.

This system will find our way in the sky, through storm and fog, sure of itself on an invisible road rolling onward and downhill within 50 feet of our landing pad. Compared with its ancestors, compared with those produced during the age of piston engines and metal fabrication now past — the newest models are roomy, efficient, fast, comfortable, and quiet. How did they gain so much in forty years?

Emerging From The Shadows

Powered flight made scant progress before the Wright brothers' success, in 1903, with fixed wings. The genius brothers possessed little money or education. Few early experimenters had much learning in what passed for "science" in aeronautics. The masses saw it as fantasy; they laughed off the news of the Wrights' first flights in North Carolina. Aviation evolved from humble beginnings.

Now we live in an age jaded by electronic wizardry; a world without helicopters is hard to imagine. Scoffers and doubters still live among us. Even Wilbur Wright: "The helicopter does with great labor what the balloon does without labor, and is not fitted for horizontal flight. It falls with deathly violence, for it cannot float nor glide. It is worthless."

The comment was in 1909 letter by the man who, with brother Orville, made history in 1903. In his way and time he was right. Knowledge of aeronautics was rudimentary. The hostile air you tried to explore could dash you to death, or smash your machine to junk. Wise men wrote of "the perils of the heights." And he who devised a machine to fly must solve **all** the problems at once. You could not fly unless you combined the basic things. The Wrights were forced to design and build their own engine.

Over three decades the helicopter slowly emerged in the shadows ("not fitted for horizontal flight," Wilbur said), in a world flying ever faster on rigid wings. The airplane brought the discovery of speed — anyway, it was more fun than the baffling problems of vertical flight; some even preferred the ornithopter concept, wings flapping like the birds and bees.

The men who invented the airplane averted the problems of rotary wings. They could not meet the demand for power, for control, nor the problems of horrendous vibrations and steady behavior, stability in flight. In a wind tunnel they could study aerodynamics of a wing — a single motion — but how do a rotor, whirling around and moving forward all at once?

Other inventors heard different music. They struggled over many decades, in all Western countries, against certain failure. They measured progress in tiny increments for payload, speed, range, and altitude; of such things we measure today in tons, thousands, hours, days. Let us consider the highlights of this colorful story.

In 1907, before Wilbur Wright's "worthless" comment, Paul Cornu and Louis Breguet, Frenchmen, each had limited success flying a copter of his own design. Breguet originated the cyclic and collective controls. In 1909 a Russian, Igor Sikorsky flew a model copter, but it lifted no more than its own weight. One year later he turned to fixed wings; in World War I he built seventy-five biplane bombers for the Czar's army — the world's first multi-engine aircraft. He would not return to rotary wings for thirty years.

Also in Russia at that time, Konstantine Antonov and an engineer named Yuriev designed and built (but never successfully flew) their own models. After the Revolution a new Soviet technical institute began working with helicopters, the start of a major industry.

The Americans Emile and Henry Berliner, father and son, built several types of helicopters in the 1920's. Marquis Pescara, an Argentine living in Spain, experimented with rotors of many blades. In a series of tests in Spain and France, he won notice for good control and for understanding the principles of autorotation.

About that time, George de Bothezat, a Russian emigre, built a complex four-rotor type under contract to the U. S. Army. He set records, hovering as long as 1 minute, 42 seconds. But he had such troubles that the Army gave up after spending more than $200,000.

There were many others, pioneers who struggled with aerodynamics, stability, control, materials and structures — problems that inventors had solved in the airplane, or at least managed. Helicopters also had severe vibrations and needed engines of high power and light weight. Stubborn men slowly wore down the problems of vertical lift.

Autogyro Invented

Juan de la Cierva made an advance in 1923 with the *autogiro,* but it was a fixed wing mutation. It had the fuselage, tail, propeller, and engine of an airplane, but a free-turning rotor in place of wings. As it moved forward its blades windmilled and produced lift. It could not fly vertically, but could take off and land with a short run on the ground.

Cierva developed the "flapping hinge;" the blades moved up and down freely on the rotor hub. That innovation relieved the problem of unequal loads on rotor systems; it cured their tendency to flip upside down out of control. Some helicopters used flapping hinges for a time.

The autogyro met quick acceptance. It carried passengers and mail, did crop work, and found many uses. In the hands of users, it boosted interest in vertical flight. As with its fixed wing ancestor, the development gained because pilots and customers worked with it in the real world in England, Italy, France, USA, and Germany.

The autogyro was a harbinger. It *ratified* the helicopter for its promise of vertical lift, and was simpler to design. It spurred interest in helicopters.

But the users found flaws in the autogyros produced in quantity over the next three decades. They had poor stability, low speed, and modest payload. Myles Ruggenberg, Scottsdale, Arizona, retired FAA inspector, who gave many autogyro flight tests in 43 years of service, growled, "Autogyros had the poor characteristics of the helicopter and none of its good characteristics. All the poor characteristics of an airplane and none of its good characteristics."

The autogyro went the way of the hybrids for that reason. Only the helicopter truly masters vertical lift.

Wilbur Wright died in 1912 but brother Orville presided over a thriving industry, and in 1934 he was writing, "The helicopter ...offers several seemingly insurmountable difficulties. Many machines designed to be lifted vertically by means of lifting screws have been proposed. As long ago as 140 years helicopters in model size were successfully flown, and yet it is only of late years that machines of this kind, built in full size, have succeeded in rising vertically for short distances....

"None of these machines has as yet even approached a stage of usefulness. Our own government, after spending more than $200,000 in building such a machine [de Bothezat's], and after surpassing all previous records for duration with it, discounted experiments because there did not seem enough promise of usefulness to justify further expenditure....

"Experiments with this kind of machine are so costly, and the chance of developing anything having a commercial value so remote, I do not think any individual can afford to undertake them as a business proposition."

Unaware or unconvinced of strides already made, Orville scoffed. Time had passed him by, this aging leader in a thriving industry. Even as he wrote, helicopter history was on the march.

French, Germans Succeed

On December 31, 1935, a coaxial-rotor design of Louis Breguet and Rene Dorand flew officially timed runs at 100 kilometers per hour (62 MPH). It had mechanical problems and carried no payload, but it performed very well, and flew thousands of feet above ground. It set records and held promise for more.

Months later the Germans Heinrich Focke and Gerd Achgelis made a copter with two rotors mounted side by side — the first practical one, with excellent stability and control. It made a famous stunt flight inside the Berlin Sports Palace, accomplished the first landings in autorotation, and set new records for altitude and endurance. The French and Germans made these strides on the eve of world war.

The war cut short the French, but Nazi Germany produced and used hundreds of copters. After Focke, Anton Flettner made "synchropters" (intermeshing rotors placed side-by-side) for Hitler's navy. But then, defeat in war ended German work in this field for years. The scene shifted to the United States.

In 1939 Haviland Platt, an inventor, and Lawrence LePage, an autogyro engineer won an Army contract for the first American copter. Their XR1 was based on the side-by-side design of Focke-Achgelis. It flew successfully but Platt-LePage starved for capital in this expensive process. After six years, the McDonnell Aircraft Company took over, but to no avail. The project was abandoned. Nor did anyone else develop and produce the side-by-side rotor system.

In the 1920's and '30's the aviation genius, Igor I. Sikorsky, now with United Aircraft, produced a line of "Flying Clippers," great flying boats which made Pan American a "world" airline. But he had not given up on helicopters. In 1928 he did some researches (perhaps drawn by autogyro success) and applied for patents. In 1939, on a modest budget of $50,000, he returned to helicopters for good.

Igor Sikorsky, in his VS300, first successful copter (he always flew with hat on.)

On September 14, 1939 Sikorsky took to the air in his VS300, with a three-blade main rotor and three tail rotors. It was not without flaws, and would not fly forward, but the inventor went on and proved his design principles valid.

With this small start, Sikorsky took the lead in the dawning helicopter world. He fostered the single main rotor as the popular design. Before the war ended Sikorsky produced 321 helicopters – the R4, R5, and R6 were the world's first turned out in large numbers – for the U. S. Army. To give him his due, Sikorsky's genius spanned three careers in a half-century.

Despite a different background, Arthur M. Young's career had parallels with that of Sikorsky. The two men spent years to make a safe, reliable, practical helicopter. Both had the vital backing of major fixed wing firms, Sikorsky at United Aircraft, Stratford, Connecticut; Young at Bell Aircraft, Niagara Falls, New York. Each man did serious work in 1929, each favored coaxial rotors to overcome torque, and both changed, in the end, to the anti-torque tail rotor.

Perhaps their differences exceeded their similarities. Sikorsky arrived in his adopted country as a poor refugee while Young came from a family of means, and a fine education in mathematics at Princeton. Young turned to the copter as "something to invent," as he said, rather than for interest in aviation. He invented one helicopter, while Sikorsky had a lifelong career in both fixed and free wings.

Young built and flew models for thirteen years to get a practical design. But he never assembled a full-size helicopter until he began work at Bell in 1941. In a temporary shop (he wrote) "I mocked up an engine, and 20 feet away, a tail rotor. How would I ever fill the space between with actual machinery that would lift 2,000 pounds?"

How Young answered that question is, of course, history. He stayed six years with Bell, on contract, to see his invention go on the lengthening assembly line. Then he turned his talents to philosophy, teaching, and writing.

"It was the loyalty and dedication" of people at Bell, the genius mathematician wrote, years later, "not the helicopter itself that I think of as the main accomplishment, for it is not making a helicopter that counts, it is the process by which it is made, and this resides in people."

Sikorsky's Work Goes On

Igor Sikorsky already had eminence as a scientist and producer in aeronautical engineering. This gentle, reserved man was moved to write of flying his VS300, "I had never been in a machine that was as pleasant to fly....It was like a dream to feel the machine lift you gently, float smoothly over one spot for indefinite periods, and move not only forward or backward but in any direction."

Sikorsky was noted for old-world manners, perseverance, and an appetite for work; he had known hardship in the Russian revolution. He loved to collect stories of rescue as the best of all things that helicopters do. He went on working past retirement in 1957. He died in 1972, at age 83.

In this field of intense rivalry, experimenters traded ideas. Young credits Sikorsky for his own decision to drop the coaxial for the tail rotor design, which wastes power. Young heard Sikorsky tell "the story of a fisherman who was paying for supplies with fish — bartering. He came to realize he was paying more in fish than he would if he used money.

" 'With the two rotors,' Sikorsky said, 'you get more complex machinery, only to avoid the tail rotor. So, it will cost you less if you pay for it with money instead of fish.' "

Young simplified his machine by mounting the engine vertically, with the transmission above and the rotor on top of that. Sikorsky visited Young's shop to look it over. Beginning with the R6, Sikorsky used that mounting arrangement.

By contract Young was to build two experimental helicopters from which Bell would make the prototype. He saw worthwhile improvements possible in a third one, but had neither authority nor money to make it. So he built it in secret, and in flight tests it was an immediate success. In fact, said Young, "that one got the company going."

Bell designated the new machine Model 47. Hardly four months later, in March 1946, the two-seater won the first government "type certificate" for commercial sale — ahead of all rivals.

Bell Sales Flop

In sales this first commercial product of Bell merely bombed. Instead of 500 Model 47's the first year as planned,

Mathematician Arthur Young invented and first flew Bell Model 30.
It became world's first commercial copter

Bell sold less than 100. The firm had to create a market for this novel contraption.

In time the Bell 47 went to war in Korea. Its distinctive bubble greenhouse cabin and long tail — known to millions on TV's "M.A.S.H." — won fame throughout the world. And it drew many imitators.

Although Young left Bell two years after production began, his friend and assistant, Bartram Kelley, stayed on. This former teacher (who had a degree in physics) worked with Young through years of research and came along to Bell, starting at $36 a week. In time Kelley rose to vice president for engineering, and he flew every model Bell made until his retirement.

The successes of Young and Kelley would not have been possible without Lawrence D. Bell — neither engineer nor pilot — who had only high school education. In 1935 he founded the company that made large numbers of fighters and bombers in wartime, and went on to produce the first American jets and the first to crack the speed of sound.

Lawrence Bell was commonly quoted: "Man wants to fly like the birds, not like a bat out of hell," and he foresaw the big future in helicopters. Even before success was assured, he dropped fixed wings, as Sikorsky had done. His employes noticed that he refused to fly in airplanes, even the airlines, after his brother died in a test flight crash. Courage was his legend, vision his guide.

With numerous modifications — twenty-two different versions — Bell made 6,263 Model 47's in twenty-seven years, more than any other to that date. About half of them went to civil use; many hundreds more were converted from the military version. There is widespread belief that Model 47 — already forty years old — will continue working beyond the turn of the Century.

It was a precarious start, but Bell (now Bell Helicopter Textron) became the world's largest producer of helicopters (7,000 Jet Rangers and over 12,000 Hueys) — some 31,000 altogether, by 1984. The company now markets a range of civil and military models such as the 222 and 214ST. All have turbine engines. Foreign factories make Bell models under license.

In recent years Bell has turned away from the two-blade rotors — which make the familiar slap-slap sound — toward four-blade types. They have better lift and lower noise levels.

The Sikorsky company concentrated more on large copters, and more for military – especially the HH3 and H60 series – than civil markets. Its S64 Skycrane, with 72-ft. rotors, designed especially for heavy lift of cargo or passengers in pods, can tote up to 20,000 pounds on a sling. It features a rear-facing seat in the cockpit so the pilot can see while he deposits the load immediately below.

Four S64's were credited in the Viet Nam war with retrieving more than 100 crashed and disabled aircraft, some from behind enemy lines. The S64 came out in a civil version in 1968. With modifications, the S64, H53, and HH3 are among the largest and heaviest in the western world.

Sikorsky today makes only two civil types, the S76 Spirit and the S70. The Spirit is a twin-engine executive chariot for 12 passengers, while the S70 is a modified military H60 Blackhawk.

To give them their due, Wilbur and Orville Wright wrote what they did from another side of genius. Wilbur died young – what more might he have contributed? To understand these men is to know that they not only changed the world, they were voices in the wilderness.

Nothing came easy, but it came. Commerce and industry, public safety, and many varied needs in war and peace, found the airplane and the helicopter – both products of imaginative men such as da Vinci, d'Amecourt, Cayley, and the rest. Now, after forty years of development, a world without helicopters is beyond imagination.

14.

The Twirly Birds

In Wright bros' wake, Igor, Breguet, Focke, Young, et al
Challenged the aero pioneers who only saw the way
On frozen wings.

Free whirling wings did balk the tamer's yoke, but dogged men
(A tiny legion) prevailed, to learn: like gold, like love,
Lift is where you find it.

Why vast real estate, and airport warehouses, for rigid wings?
Free-flying copters rove earth's heights and go,
From anywhere to land wherever.

As World War II was mighty aviation's war, its few helicopters stirred intense interest. They enticed inventors, even from outside the field, to study the problems of rotary-wing flight.

In 1943 a small band of engineers, scientists, pilots, and true believers formed the American Helicopter Society; it grew steadily and attracted a foreign membership. The Helicopter Association of America (HAA), made up of users, pilots, and owners followed in 1948. The Society fosters research and technology, and the Association promotes standards for the business and for safety of flight.

The industry grew slowly, as helicopters sought a foothold in the established world of fixed wings. Some of the progress made with airplanes, notably power plants, aided the helicopter, but airplanes with their glamorous speed drew greater attention.

Today the Association accounts for almost 10,000 civilian

copters in the United States, 29,564 licensed pilots, 4,232 heliports and landing pads, and over 2,700 operators who log 3 million flying hours per year. The 1984 world total of civilian helicopters, says the authoritative *INTERAVIA* magazine, was 15,803 — nearly 3,000 more than the total of turboprop and jet airplanes used in business.

Members of the Association achieved a low accident rate, less than twenty per hundred thousand flying hours. "Safety in all aspects of helicopter operations has been our principal objective," says Charles Johnson, Association president. "We foster high standards of professional and ethical conduct, and promote the helicopter as a safe, practical, reliable, and cost-effective form of transportation."

To know its history is to see that neither design genius nor riches were enough to make this machine practical and popular. The pilots, engineer-mechanics, and users contributed the rest of it — as we detailed in previous chapters. Among these, the "Twirly Birds" at first meant those who had flown solo in a helicopter before August 14, 1945, the day the great war ended.

Allen Price, engineer and pilot for Platt-LePage, and former Twirly Bird president, recalls a summons from Les Morris, chief pilot of Sikorsky, to a meeting February 9, 1946, in New York City, to form this elite group.

"It was attended by about two dozen candidates. Igor Sikorsky ran the meeting, proposed the bylaws, and so on....In addition to Jimmy Viner and other Sikorsky pilots, and Buck Miller and I, there were Captain Frank Erickson, Stew Graham, Steve Tremper, and Bill Knapp from the Coast Guard.

"Sikorsky told some of his early adventures with the VS300, flown by himself, Mike, and Serge Gluhareff. Charles Lindbergh was there."

Price added, "Frank Erickson trained Coast Guard pilots at Floyd Bennett Field using a simulator improvised from a traveling overhead crane in a hangar, with a cockpit hung on a pantagraph; hovering, circles, even figure-8's were possible. They built a ship-motion simulator, a 20 by 20-ft. platform which rocked for practice landings aboard a ship tossing in heavy seas. They dubbed it 'SS Mal de Mer.' "

Many of these pioneers were either self-taught or former autogyro pilots, Price believes. In later years the Twirly Birds eased the membership criterion for later solo flight dates. They realized that all had made their mark in this brave new world.

Igor Sikorsky had the first copter pilot license, and Frank Piasecki, No. 2. Third was Floyd Carlson, famous engineering test pilot for Bell. He was among the earliest of this small elite. Only a few hundred qualified in helicopters before war's end. The Twirly Birds were those who made the faltering new device work.

Piasecki Made Strides

As Europe was badly ravaged by war, the initiative for copters shifted to the United States. Besides Igor Sikorsky and Arthur Young, Frank Piasecki made major strides at Philadelphia.

With a group of students from the University of Pennsylvania, Piasecki designed a single-main-rotor model; it was the second successful U. S. helicopter. He then formed Piasecki Helicopter Corporation and produced tandem-rotor machines for the armed forces. His career rose fast.

While Sikorsky and Young turned from twin rotors to the single, Piasecki went the other way. For the U. S. Navy he built the first fore-and-aft rotor, ten-passenger HRP1 "Dogship." The sailors dubbed it *Flying Banana* for its fuselage bent in the middle. (That shape was necessary to make its rear rotor turn clear above the front one.)

Insiders say Piasecki began with a side-by-side design, but it would not fly forward, so he moved the cockpit to one end and made it go sideways, in effect, with the rotors in tandem. Whether that tale is true or not, Piasecki produced many tandems for military forces, and achieved real success. In 1952 he made his largest, the experimental USAF H16, with 82-ft. rotors and gross weight of 32,000 pounds. It cruised at 166 MPH and carried as much as seven tons.

In 1955 Piasecki's firm was bought up to become the Vertol Division of Boeing Airplane Company. A civil version of the military Vertol CH46 tandem, Model 107, flew passengers around New York City for some years, and has since seen extensive use in heavy lift.

Kawasaki Industries still makes the 107 in Japan under license. Vertol today markets the Model 234 (military CH47) for heavy lift and oil platform support. It carries forty-four passengers and a crew of three, the largest in the West, with a price tag over $15 million.

In a new firm, Piasecki Aircraft Company, the inventor continued with other vertical lift projects, such as a "Heli-Stat Heavy Vertical Airlifter." This combines four helicopters with a blimp which is expected to lift most of its 24-ton payload. The four copters together will provide the rest of the lift and the means of control.

Dozens of other American aircraft makers turned to helicopters after World War II, among them:

* Hughes Aircraft took over a helicopter design from Kellett Autogyro Company, but then developed its own single main rotor models. Hughes became the third largest manufacturer.
* McDonnell Aircraft, St. Louis, took over the Platt-LePage twin-rotor [see chapter 13] then dropped out of helicopters. In 1984 McDonnell-Douglas bought out the Hughes Helicopter Division.
* Lockheed Aircraft did pioneering work with rigid rotors for the Army, but gave up after a few years.
* Gates Learjet entered the field in 1969 with a $450,000 high tech "Twin Jet". But recession nipped it before a prototype was built.
* Still another airplane firm, Cessna, in 1956 marketed a neat high-performance, four-seat "Skyhook," designed by Charles Seibel. For a demonstration it landed on top of 14,110-ft. Pikes Peak in Colorado. But promise fell short. As Cessna sold only twenty-three Skyhooks by 1963 (at $80,000 each), the company bought them all back, scrapped them, and departed the field.

Numerous rotorcraft builders have come and gone; some historians estimate 500 of them in fifty years. The failures trace, indirectly, at least, to the lack of government aid and support for this civilian endeavor. The record shows large sums in development for warfare, but little for civil uses, or facilities or navigation aids.

Helicopters of two to five seats have been the most popular in this recital of success and failure. Stanley Hiller, a versatile inventor, tried a number of novel design ideas, including a one-seater powered with ramjet tips. It "made a noise like a banshee, and, in autorotation, virtually fell out of the sky." So said witnesses.

In the end, Hiller produced a sturdy two-place single rotor helicopter; Model 12 became the workhorse of agriculture and utility. The Hiller company has changed hands several times in recent years; Model 12 is still in production.

Other American makers include Kaman, Enstrom, Spitfire, Texas, Brantly, Umbaugh, and Rotorway. For the U. S. Navy, the first Kamans were *synchropters,*[1] the only ones made with intermeshing rotors.

Europeans Get Busy

Europeans got busy as they recovered from the war. Germany, France, Italy, and England broke into the U. S. monopoly. The French marketed helicopters in many nations; they built a plant in Texas and won a large contract for the U. S. Coast Guard H65.

At first the French set out to improve on the work of the former (German) enemy. Nationalized factories vied with each other in a series of new ideas, and they early made strides with three record-setters, the Alouette, the Lama, and the Djinn.

Alouette copters won a reputation for reliability and good performance. Alouette II was the first ever produced with gas turbine engine. The seven-seat Alouette III, made in France in 1975-79, remains in production in India and Romania.

The 560-HP Lama, designed for high altitudes, first flew in 1969. The French test pilot, Jean Boulet,[2] set an altitude record of 40,820 feet, which still stands.

Boulet stripped the Lama of extra weight, even the battery and starter, but when he reached to the top of his climb and cut power for the glide back, the engine stopped dead. "Without a way to restart," he laughed, "I had to make the longest autorotation in history." The Lama is still in production.

Starting in 1957 the Djinn won honors, the only production copter with jet-tip power. In this light two-seater, a gas turbine engine supplied compressed air to its rotor blade tips. It had no torque, no danger nor waste of fuel and power on a tail rotor.

The Djinn saw use in crop work and in the French army. It had light weight and simple upkeep, but gobbled fuel, would not exceed 75 MPH, and had poor control in hover. Production ended in three years.

At least a dozen other jet-tip models were built in the USA and western Europe. The British flight-tested a hybrid jet-tip "Fairey Rotodyne," for forty passengers. It had two

[1] See Appendix.

[2] Boulet is author of "Helicopter History" published 1983.

engines, one large rotor, and two propellers on wings. It took off and landed as a helicopter and flew forward as an airplane. Disappointing flight results caused the government to dump the costly program, however.

The American jet-tip Hughes XH17, weighing 44,000 pounds, had a 130-ft. two-blade rotor for lift as a flying crane. It too ended at the flight-test stage.

Today, the nationalized French company, the Helicopter Division of Aerospatiale, turns out high-tech helicopters ranging from the Lama to the nineteen-seat three-engine Super Frelon. The French were first with gas turbine engines; they also led in using "high tech" composite materials; their rotors have "unlimited" life, (not replaced after any specified period of use).

In Germany, the old Messerschmitt firm linked with Blohm and Boelkow (as "MBB") to produce an advanced seven-seat Model BO105. It is sold in many countries. The BO105 has two engines and a four-blade rigid rotor. ("Rigid" rotors are flexible enough to permit twisting and bending as the blades turn in rotation.) MBB has joined with Kawasaki to produce the "BK117 Space Ship," also with rigid rotors; it seats up to eleven persons.

Under the Common Market, cooperation brought together the former enemies of western Europe in many ventures. France and Germany (MBB and Aerospatiale) have formed a partnership for a new Army attack helicopter, with different versions for each country; it is planned to be operational by 1993. The British and Italians also are joined in military production.

The full story of helicopters behind the Iron Curtain is not well known, but native Russians have done as well as those who emigrated. They have set a share of world records. One of those — climbing to 6,000 meters (19,690 feet) in 4 minutes 47 seconds — was set in 1983 by two women pilots, Tatyana Zuyeva and Nadezhda Yeremina, in a Kamov 32, which has a three-blade coaxial rotor system.

Most of the estimated 24,000 helicopters in the USSR were the work of Mikhail Mil, who is said to be what Mikoyan is to fighter planes. They come in a variety of types and sizes; about 12,000 are in civil use, the rest in military garb. The Soviets' seven-blade Mi6, lifting 26,000-pound internal loads, is the world's largest. The Mi24 is believed the most heavily armed.

Under Soviet license, PZL (Polski Zaktady Lotnicze)

makes the Mi1 "Bazant," (Pheasant) at Swidnik, near Warsaw, Poland. It is sold in many parts of the West as well as in Warsaw Pact nations, especially for farm work.

Caution: High Stakes

The twists and surprises that stud this story from earliest days must give pause to one who would guess the future. It has ever been so in aeronautics, where big business took over from the striving pioneers as the utility of flight grew and expanded. They made it profitable, as indeed someone must.

With such big business and high stakes, however, came a measure of caution — not the daring of LePage and Breguet, Sikorsky and Focke, Hughes, Piasecki, Bell, and Hiller. For any others waiting offstage today, the stakes are far higher.

The Sikorsky firm founded by the genius from Kiev continues in the lead after his death. As he adapted to change in three countries and three different careers in aeronautics, the Sikorsky Division of United Technologies Corporation goes on under younger successors.

For the future, Sikorsky has put money and effort into several promising ideas, notably the ABC ("Advancing Blade Concept"). It has rigid rotors, one above the other, turning in opposite directions. This design Igor Sikorsky himself abandoned four decades earlier.

In the ABC, one blade is "unloaded" (i.e., not lifting) as it turns backward in the cycle of rotation while the other blade turns forward. The design permits higher speeds than the single main rotor. The ABC has a conventional airplane tail and two jet engines for forward thrust. In tests the ABC has topped 300 MPH, and is expected to do at least 350.

Even as it produces military rotorcraft on multi-billion dollar contracts, Sikorsky seeks to sell this proposal to the armed forces for development. Where next? Igor's spirit is under challenge.

The spirit of Lawrence Bell haunts the company that bears his name, as it too works on military contracts. Bell has turned on a course away from the tail-rotor helicopter. The company is betting high on the so-called *tilt rotor,* a hybrid "XV15." In a joint effort with Boeing Vertol, it vies for a large military contract that would pay off the expensive process of development.

The XV15 is a product of large amounts of government support, over thirty years of research (and failure), but its prospects now look bright. Basically, it is an airplane with two 25-ft. rotors, at the tips of stubby wings. It takes off and lands as a helicopter; its rotors tilt down into propeller position to fly forward.

The XV15 is said to perform well in all modes, and has reached 350 MPH. Among other things, the design requires a heavy structure to support rotors side by side. It cannot land with rotors down, in airplane propeller position, as the blades would strike the runway. It is clear the XV15 must be a first-rate helicopter; aviation experts see no need for another airplane of medium size and high complexity.

Decades after the helicopter's birth, one wonders that the old past echoes in the fixed wing world, with its ponderous airports, choked airways in the sky, and gross airline terminals of warehouse size. That world grew its own barnacles, its zealots who disdain the complicated helicopter that the Wright brothers spurned. Do they really understand it?

The helicopter professional and trade groups never found the government support nor research as still goes to develop the airplane. "It took major government subsidy," says John Zugschwert, executive director of the American Helicopter Society, "to build and maintain the enormous complex — the airways, airports and their accouterments devoted to airplane traffic — that we have today.

"We have no such facilities to bring the benefits of the helicopter to our citizens. Whether they will be built remains to be seen but the growth of the helicopter industry over the past four decades, together with the progess under way now, all but guarantees that it will be done. It must be done."

Zugschwert cited the example of the Washington National Airport, where big jets frequently take off and land over densely populated areas. The 1981 Air Florida crash into a mid-town bridge raised public outcry over the safety of that airport, and he offered a bold solution — make it "National Heliport."

"We would eliminate the noise footprint of the *long* fixed-wing approaches and departures. It's not as far-fetched as it sounds."

Zugschwert thus fingered both the fierce airport controversy and the narrow mind-set of the capital that readily accepts the helicopter in the White House yard, but none others. Helicopters never use the heliport on the Department

Sikorsky ABC, fast new coaxial rotor model under development.

XV15 being developed by Bell — side-by-side rotors do helicopter takeoff and landing, then tilt downward in forward flight.

of Transportation building roof. Even the local police copters are forced to fly from National Airport — located across the Potomac in Virginia. It's a story of frustration.

Probably typical of the Washington attitude is the Smithsonian National Air Museum, with elaborate displays of fixed wing and space flight artifacts, but only minor ones of helicopters — placed in a small room labeled, "Vertical Flight." Arthur Young's historic Model 30 is hidden away in dry storage. And a popular movie about flight — although liberally filmed with helicopters — contains no picture nor word about them. It is all fixed wing.

Civil Needs Ignored

Even more critical comments were voiced by Joseph Mashman, after forty years in helicopters. The retired vice president of the Bell company complained of "indifference to the needs of civil users" by the FAA and NASA, the "space agency." He charged that none of NASA's goals hold "....any promise of significantly affecting civilian helicopters in economics and usefulness." (NASA has the same mandate for research on aeronautics as for space activities.)

At the "Annual Forum" of the American Helicopter Society, he noted, NASA's Deputy Administrator Hans Mark, along with military officers, boasted that the XV15 tilt rotor would enhance civilian development. Mark's contribution of wisdom" said Mashman, "was to assert that by the turn of the century the helicopter would be extinct and replaced by vertical-lift jet airplanes."

The tilt rotor, Mashman asserted, "will have higher acquisition cost and less useful load than the pure helicopter. Its cost-effectiveness will pay off only for long distance travel where its higher speed and lower fuel consumption provide a tradeoff."

"The military tilt rotor proposal," he went on, "has very little potential for use as a civil aircraft. Like the Hughes AH64 attack helicopter, it is designed to operate in hostile enemy environment." It has low payload, high running cost, and advanced equipment not needed for civil use.

The Personal Helicopter

In the past thirty years the pace of civil airplane development has slowed markedly. Aeronautical scientists say the airplane has less potential for basic improvement than the helicopter. Both are costly, compared to other vehicles, if they can be compared.

The copter costs twice as much as any equivalent airplane — due of course to smaller production and lack of convenient facilities. But the helicopter has the potential to supersede the automobile, as it can give convenient personal transportation like the private car. It is the only aircraft you can take home.

The Sikorskys and Youngs and Piaseckis of the future, like da Vincis and d'Amecourts of the past, can do a service for humanity by reducing costs.

Frank Robinson has such purpose with his R22 helicopter. An old friend called him "single-minded, unable to compromise — his copter is more important to him than life." In his small factory at Torrance, California, Robinson dominates all, much like Sikorsky, Bell, Piasecki, and others of the past. He tends to the testing, engineering, production, and sales like a mother, they say, and takes a personal hand with every machine on the production line.

He describes himself as "committed," and he learned by working for six different helicopter makers. He's a native of Whidbey Island, Washington, where airplanes and helicopters

Robinson R22, newest, smallest commercial helicopter.

commonly cross the sky. He has an engineering degree from the University of Washington.

Robinson struck out in 1973 to market the simplest helicopter for personal use, for $30,000, lower than any other by half. The process of development took three and a half years, at least $2 million, and countless engineering and flight hours to get FAA certification. It was the first new small copter in twenty years.

Robinson's R22 emerged with more speed and performance than any comparable, but still the smallest civil production copter ever made. Finally on the market in 1979, the R22's price had already inflated above $40,000, and continued growing steadily.

There was product liability; no new machine comes without flaws, and Robinson ran into modifications and fixes. "That's where the big costs come from," he said, "for the lawyers and the litigation — in fact, this problem threatens all of general aviation today."

Again, "though we meant it only as a personal helicopter to go from place to place, we found it was the most popular trainer on the market. That meant rough use by the untrained, the non-professional pilots. We soon had accidents, and one accident is too many. If we couldn't get the accident rate down I would have quit."

That involved Robinson himself in the techniques of flying, the fine points of piloting as, "We had to know how they make the errors that cause accidents."

To reduce the accident rate, Robinson set up a novel low-rate insurance plan for pilots and instructors who take flight training at his plant. Although some 450 Model R22's now fly in twenty-three countries, accidents have declined, Robinson said. More than 400 instructors have taken the course. He claims "the lowest accident rate in the industry."

The R22 has been making a good reputation for training and light cross-country travel. It carries pilot and passenger in comfort at 110 MPH. Its 25-ft. rotor runs at low noise level and enables flight and landing in tight corners. The maker believes its best atttributes are reliability, performance, and simplicity.

Though priced at the bottom of the pack, the R22 is not quite cheap. In 1984, the price tag had reached $81,850. Robinson blames inflation for at least sixty per cent of that rise, but "people demanded higher performance, more advanced design, so we added things. But we have kept it as simple as possible."

In running costs, the R22 compares well with fixed wings — with autos too. Robinson claims it costs less than $50 an hour to fly, about 50 cents a mile. Will costs of buying or flying drop? He sees no possibility in the near future. All costs in manufacturing go on rising, year by year, and there is liability, a factor which hardly troubled the early producers.

Who knows what more may develop? A full decade after starting in this volatile business, Robinson sees his sales and production on the rise, and has plans for bigger things. He is only 54, committed to his product — convinced and enthusiastic as were the inventors and the Twirly Birds who created the helicopter world.

The modern helicopter has gone far from the days of barnstorming airplanes, airmail service, and the autogyro, but has not lost the dreams of Sikorsky, Bell, Young, Breguet, Piasecki, LePage, nor the other ancestors of its long history.

The Twirly Birds, every last one, swear that a helicopter is the safest and the most fun to fly. They know the best takes a little longer to perfect. To be sure, lift is where you find it.

Appendix:

How It Works

How do "heelicopters" fly? They're wild machinery — the Earth naturally repels them.

— A bystander

Engineers tell a different story. As safety always comes first in flying, we begin with *autorotation,* the classic safe-landing feature of helicopters.

In autorotation flight, the pilot keeps the rotor turning so it develops lift — the basic essential of flight. He lowers the pitch of the rotor blades, which makes the copter move downward and the airstream flow upward. This moving air makes the rotors spin exactly as it turns the windmill. As the copter glides to earth, the pilot stops to hover just off the ground. He lands from that position, with forward motion stopped.

As the word "auto-rotation" indicates, the blades (wings) of the helicopter rotate automatically, without the driving engine. Under full pilot control, the copter glides like the fixed wing airplane, but autorotation ends with a landing at zero speed, on any spot large enough to clear the rotors.

The autogyro, a primitive aircraft, flies by autorotation; as its propeller pulls forward, the rotor blades keep turning. A 'gyro can only take off like an airplane, although it can hover briefly as the helicopter does in landing.

In helicopter autorotation, the instant the engine power slows or stops, the rotor disconnects through a "free wheeling" clutch. In *normal* flight, the copter glides down (with engine power reduced) to hovering position close to the surface. It then lands, with or without power, but no forward movement. This is why the helicopter needs no runway, no long approach path free of obstacles, no wheels nor brakes, as it stops before touching down.

Autorotation takes piloting skill, but in emergency, it averts the high-speed crash impact that airplanes must risk in landing. Although nothing that moves can achieve perfect safety, autorotating helicopters reduce the worst of damage and injury. This helps to explain why copter accident rates compare well with other vehicles.

Four Forces of Flight

There are those who claim the helicopter is a push-pull up-down-type gadget. Maybe so. Its unique flying qualities may look simple, but they come from complex aerodynamic forces.

The free-flying qualities of the copter derive from *lift* of the Rotor Disk. We see lift in the flight of birds and insects, in wind blowing flags and trees. Like the wings of an airplane, rotor blades are airfoils shaped to make lift.

Three forces combine with *lift* to make powered flight possible: gravity, thrust, and drag. Everybody knows *gravity,* the opposite of lift that pulls us to Earth. Engine *thrust* turns the blade-airfoils (like a fan) so they create lift. *Drag*, resistance of the air, is the backward pull on anything (viz., airfoil) in the windstream. Altogether, these four forces provide the means of control.

Helicopter = Total Freedom

In our three-dimension world only the helicopter moves with total freedom. It travels in all directions — backward, up, down, sideways, spiral — and stops anywhere.

In engineering terms, it moves in *two different modes on three axes of rotation.* It moves in a straight line and turns, at the same time, around its vertical, horizontal, and lateral axes, or centers of rotation. That adds up to *six different motions.* It brings us flight more like the bee and the hummingbird.

The vertical axis is a line through the copter center of gravity (CG) from top to bottom. The horizontal axis runs fore and aft through the CG, the lateral axis likewise from side to side.

Thus the copter can fly all these motions:

+ Up and down, like an elevator;
+ Yaw (pivot) and turn to either side;
+ Forward and backward;
+ Roll (bank) from side to side;
+ Slide (skid) straight sideways;
+ Pitch nose up and down.

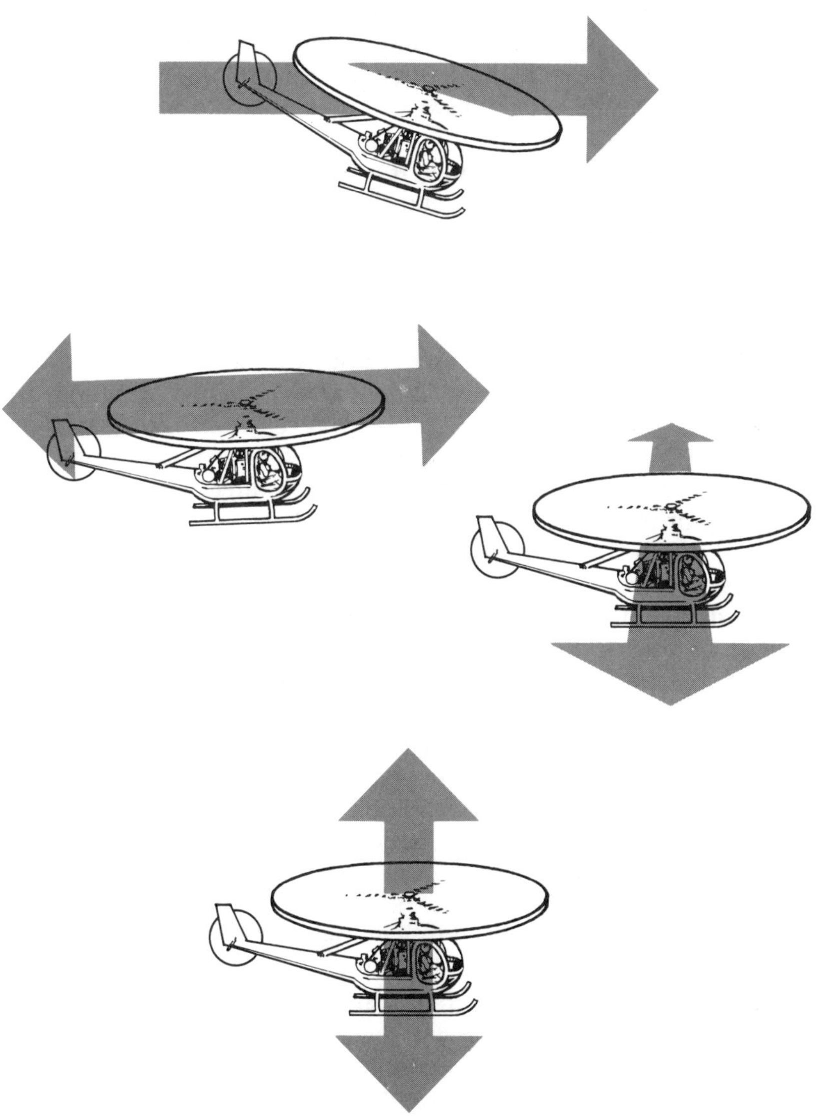

That Live Disk

Think of the rotor as a *disk* rather than a set of airfoil blades; those blades produce lift only when whirling in a disk anyway. The rotor disk – equivalent to the airplane's wings – is what the controls manipulate, what the pilot flies, what makes the copter go where it is wanted, what engineers use to calculate its functions.

Lift always works at right angle to the disk, therefore, when the disk tilts, a part of that lift makes it move toward its low side. As with the frisbee (another flying disk), the more it tilts, the faster it moves. The amount of lift derives from:

1. the speed of rotation, in revolutions per minute (RPM)
2. blade length, width, and shape
3. angle between the blade and the wind
4. engine power
5. density of the air. (Warmer, less dense air at higher altitudes give less lift.)

Change in any of these factors directly affects lift. Therefore, we control the disk by forcing change or acting against it. Thus, with the *cyclic pitch lever,* main right hand control, the pilot tilts the disk to move forward or in some other direction. The copter then flies ahead, sideways, angling or to the rear, in the direction of tilt.

With the *collective pitch lever,* at his left, the pilot raises or lowers pitch of all rotor *blades* at once. By thus changing their angle with the airstream, the disk produces more or less *lift.* This is the up-and-down-lever.

In certain copters the *engine throttle* connects to collective pitch. The pilot twists it like a motorcycle grip to get more or less power. On gas turbine engines, a governor regulates power on a separate control.

Although the *foot pedals* look like fixed wing rudder controls, the helicopter has no rudder. On most models the pedals change the thrust of a tail rotor; they swing the nose to one side or the other.

Rotor controls respond to light pressures; the pilot does not move hands and feet. With large inertial forces and a high-powered engine, the live disk reacts fast, giving the copter its agility and quickness.

Of course, all controls and maneuvers have limitations, and the helicopter flies in an environment of shifting wind velocity, gusts, temperature, and density. As the disk differs from fixed wings, wheels, tracks, propellers, and boat hulls, it makes the helicopter perform unlike any other vehicle.

At takeoff the copter may appear to head down like a bronco while surging upward. Or it may rear back, nose high, for sudden stop in the air. From a hover it seems to dive while heading upward. Again, it will move low and slow over the surface in *hover-taxying.* Such agility is all due to that live disk.

Normally, flight begins facing the wind in brief hover, then climbs steeper as forward speed increases (more lift). In landings, the pilot does the reverse — he descends and slows, into the wind, so as to stop close to the surface, then touches down lightly.

The airplane must fly forward, always forward, to keep its fixed wings lifting. This makes the fixed wing relatively ponderous, if not dangerous. Hence the airport has to be a giant industrial plant along with vast spaces reserved in the sky above.

For regular operation, a helicopter needs enough room to fly or sit down — on an ice floe, a mountainside or crater, a roof or pier or ship's deck, the bed of a truck, water or mud or grass or sand bar — or even to hold still in hover, high or low.

Engines

In recent years the turbine (jet) engine has replaced the piston engine in all but the smallest helicopters, a change more widespread than in fixed wings. Although it costs more to make, the turbine has few moving parts to wear, is more reliable and free of vibration, takes less care, and weighs less. One 317-horsepower engine weighs only 158 pounds; another weighs 434 pounds and delivers 2,500 HP.

Some helicopters have two, three, or even four turbines. They are inter-connected to all rotors. Design and materials improvements have raised the performance and life span of engines. The latest ones compare well with autos of equivalent power, in miles per gallon and cost of operation. At the same time they give greater performance, speed, and usefulness.

Torque

Helicopters have two rotors, mounted in various positions. The most familiar type *(A)* combines a main rotor with a tail rotor which is mounted at right angle to the main rotor, in order to counteract its *torque,* or twisting force. Without the tail rotor, the fuselage (body) of the helicopter would whirl opposite to the main rotor.

Unwanted but unavoidable, torque is useful for turns. Pressure on the left foot pedal reduces tail rotor thrust to the left, which swings the tail to the right and the nose left. The right pedal so controls right turns.

The helicopter turns like the fixed wing, i.e., it banks the disk right or left, and the tail rotor balances the turning force in the airstream. Unless the turn is balanced with the rudder, an airplane will "skid" dangerously, but the copter will merely move sideways. The pilot uses tail rotor thrust *against* the turn to fly sideways.

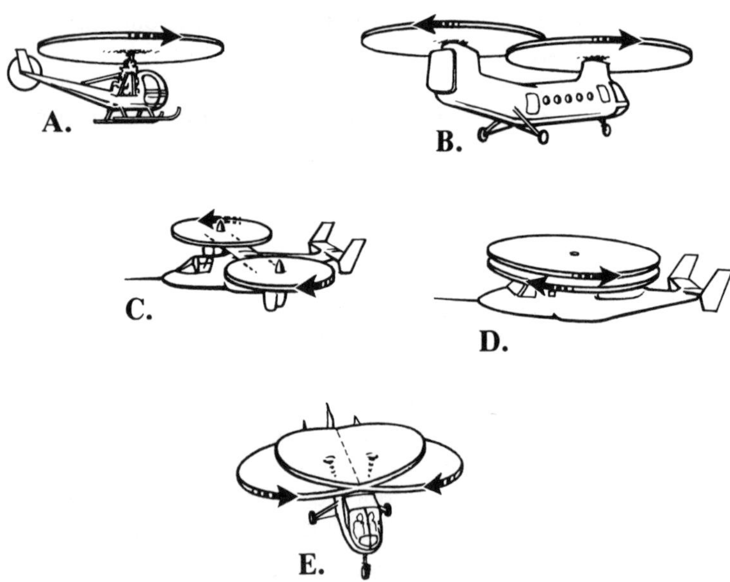

Twin rotors turn in opposite directions to cancel the torque, one of the other, so they need no tail rotor. With rotors mounted fore-and-aft *B* or side-by-side *C,* the one rotor tilts to the right, the other left, to balance the turning forces.

The coaxial type *D* with one rotor mounted above the other on a single shaft, and the intermeshing, or *synchropter, E* with two shafts, make turns by changing blade pitch (lift) on the opposite sides of the two disks.

These control functions involve complexity, hence the twin-rotor design costs more to produce, so the tail rotor type is more popular. All designs have advantages and disadvantages.

The tail rotor wastes power, requires care, involves a lengthy drive shaft, is awkward in use and a danger on the ground. Certain helicopters have large tail fins to overcome torque at high forward speeds, thereby reducing tail rotor power waste.

The "Fenestron," an example of this type, has blades mounted within a tail fin. Another concept, "Notar" (no tail rotor), uses an air jet from the engine for turn control, but this, too, wastes power.

These few of the many innovations on helicopter design exemplify simplicity and reduced weight. The so-called "Third Generation" models are made of highly durable composite materials such as Kevlar, Fibreglass and carbon filaments. Some rotor heads and blades even have unlimited life span — they don't have to be replaced at a specified time.

Limitations

Compared with the fixed wing, the helicopter has the speed of a horse, but it beats the turtle, and nobody forsees a way to make more than moderate improvement in speed. With its agility it needs no great speed to be useful, but even a modest increase could extend its range.

Speed is one element of the equation of travel, the singular reason why fixed wing air lines dominate long distance transport. And so, adding only 50 MPH could extend the practical round-trip range of the helicopter beyond its present 600 miles or so.

The speed of rotor blade tips limits true high speed. Hovering in calm air the blade tips turn at 400 MPH or more. A moving copter *adds* forward speed as the tip swings forward; as it turns back on the other side of the disk the forward speed *subtracts*. The higher the forward speed, the greater the imbalance between the two sides. This is what holds down forward speeds.

Helicopters also carry a heavy burden of drag, due to the rotor system, landing skids, tail rotor, and other protrusions, and these all increase with speed. Raising the speed by half multiplies drag by a factor of four.

Practically speaking, helicopters cannot attain speeds much higher than 300 MPH, and few of them today top 200. An unforeseen discovery could change that, however. Aeronautical science has overcome many hurdles in its colorful history. In the fixed wing, it conquered what was once feared as the "sonic barrier," to make supersonic flight virtually commonplace. It never was a true barrier.

The helicopter may appear simple, even magical, but it is only a mechanical device. The laws of physics, the interaction between its parts with the fuselage and the environment, make it one of the most complex of modern vehicles. It demands piloting skill, not because it is hard to control but because it normally works at low altitudes, in environments and confined spaces where hazards are high.

In proportion to its size, no other vehicle has such varied dynamic systems — a pair of rotors 30-75 feet or more in diameter, plus powerful engines, plus one or more transmissions and assorted gear boxes and devices. The total is a mass of inertia and vibration.

The transmission — itself a heavy item of machinery — includes a free-wheeling clutch and, sometimes, drive belts. These multiply in a multi-engine machine. One tandem model has five inter-connected transmissions. Some have drive shafts over 50 feet long, plus other devices for control. Finally, there are hydraulic, electrical, and pneumatic devices, plus an array of electronic systems.

All these factors together result in expensive maintenance on top of high production costs. The helicopter remains among the most costly of vehicles to buy and use. But there is that utility and versatility. Measured against the work it can do, it sometimes becomes the fastest, if not the only means to carry out a given task.

Bonuses — It's Fun

As noted, lift increases with speed; therefore, a moving helicopter obtains extra lift. That is, the disk produces more lift in forward motion than in hover. Unique to the helicopter, this *translational lift* permits flight with good performance and lower power in hot air or high altitudes.

It may fly forward when it cannot hover, or take off with more load in a strong wind than in still air. A wind as low as 20 MPH develops substantial translational lift.

Normally, it takes more power to hover than to fly into the wind. But the hovering copter has another bonus from *ground effect.* Rotors move large quantities of air at high speed. As this downwash air can not escape fast enough from under the disk, it forms a cushion, as thick as 30 feet, on which the copter floats with low power and fuel consumption.

It's like a big pillow. It makes the helicopter fun to fly.

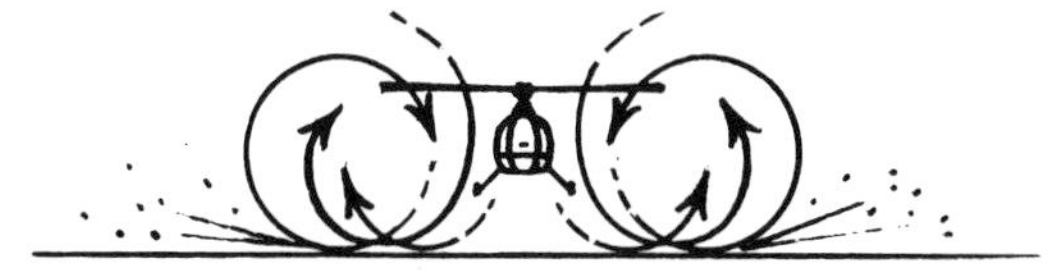

Vapor trails blade tips of Aerospatiale Puma.

Scotland Yard's Bell 222 over heart of London.

INDEX

The Author

Joe Stein logged more than 4,500 hours as a career pilot in civil and naval aviation. Since 1946 he has written news, science, and aeronautics for newspapers and magazines. He was aviation editor and pilot-reporter of the Oregon Journal, flying the first news helicopter. For 17 years he wrote in public affairs at the NASA Headquarters, Washington D.C., and was deputy director of public information. He spent six years in travel and research for this book, and is a member of the American Helicopter Society, Aviation Writers Association, and Helicopter Association International. He lives near Mount Hood, in the Oregon Cascades.

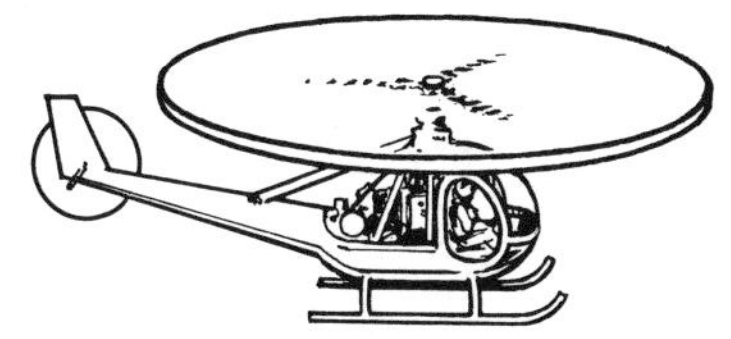